The Principles Of Cancer And Tumor Formation

William Roger Williams

THE PRINCIPLES

OF

CANCER AND TUMOUR FORMATION.

BY

W. ROGER WILLIAMS, F.R.C.S.,

SURGICAL REGISTRAR TO THE MIDDLESEX HOSPITAL

SURGEON TO THE WESTERN GENERAL DISPENSARY.

London:

JOHN BALE & SONS, 87-89, GT. TITCHFIELD STREET,

OXFORD STREET, W.

—

1888.

TO

DR. E. HEADLAM GREENHOW, F.R.S.,

CONSULTING PHYSICIAN TO THE MIDDLESEX HOSPITAL,

THIS BOOK IS DEDICATED

AS

A TOKEN OF RESPECT AND ADMIRATION

BY

THE AUTHOR.

PREFACE.

A brief sketch of the doctrine developed in this book, was first published in my work "On the Influence of Sex in Disease," (pp. 10-11), issued by Messrs. Churchill, in November, 1885.

The chapters on Growth and Reproduction were published in pamphlet form in April and May, 1886.

In November of the same year, I set forth the whole subject in an essay on "Vegetable Tumours in relation to Bud Formation," read before the Pathological Society of London.

It is necessary to mention these facts, because in a work just published, Mr. Pitfield Mitchell has advanced views very similar to mine, without being aware of my previous publications.

W. ROGER WILLIAMS.

LONDON,

April 26, 1888.

CONTENTS.

LIST OF ILLUSTRATIONS.

LIST OF ILLUSTRATIONS.

INTRODUCTORY.

THE present work is intended as an introduction to a contemplated Treatise on the Pathology and Treatment of Cancer and Tumour Formation, of which it forms the first part. The author proposes to complete this Treatise in six parts. The second part will be devoted to General Pathology and Treatment; and the four succeeding parts, to a series of monographs on the Pathology and Treatment of the chief Local Varieties of the disease.

THE PRINCIPLES

OF

CANCER AND TUMOUR FORMATION.

CHAPTER I.

GROWTH.

"The most important truths in Natural Science are discovered, neither by the mere analysis of philosophical ideas, nor by simple experience, but by *reflective experience*, which distinguishes the essential from the accidental in the phenomena observed, and thus finds principles from which many experiences can be derived. This is more than mere experience: it is, so to speak, philosophical experience."

JOHANNES MÜLLER.

INASMUCH as Pathology is but a branch of Biology, it necessarily follows that when new facts and principles are discovered in the latter, they re-act powerfully on the former.

The history of Medicine amply illustrates this statement. Thus, we are indebted to the genius of the vegetable morphologist Schleiden, for the famous

Cell Theory, which must be ranked among the most important steps by which the science of biology has ever been advanced. Every great pathological epoch has in fact been foreshadowed by some such precursory innovations.

It is astonishing what relatively long periods often elapse before the significance of such discoveries is duly appreciated. The *vis inertiæ* opposed to all progress is immense. We see this in the case of the modern doctrine of evolution. Already more than a quarter of a century has elapsed since the publication of Darwin's great work on the "Origin of Species," which has revolutionised the rest of biology; but with regard to pathology, it has as yet produced no commensurate results.

My endeavour in this book will be to contribute something towards the building up and developing of pathology, in accordance with these neglected principles—in short, to give a scientific basis to our pathology of cancer and tumour formation.

History shows that in the application of general principles there are always some anomalies at first, which are afterwards cleared up. And so no doubt it will be in this case.

In such an exceedingly complex science as biology, accurate and diligent empirical observation, though the best of things as far as it goes, will not take us very far, unless at the same time we seek out laws to bind together the immense accumulation of facts.

Empiricism of itself is but a stepping-stone to real knowledge. Those who think that the mere heaping up of facts constitutes science are certainly in error. We ought to remember what Aristotle long ago pointed out, that intellect (*νοῦς*) is the beginning and end of science.

The need for generalisation never was more urgent than at present, when innumerable guideless investigators, tempted by the facilities offered by improved microscopes, &c., threaten to stifle the truly scientific spirit under masses of un-unified facts.

When we come to examine the linked chain of living things in the light of the doctrine of evolution, we are insensibly led to the important conclusion that life must necessarily have preceded organisation; whence it follows that function—which in its widest sense denotes the totality of all vital actions—in all cases goes before structure, and is from beginning to end its determining cause.

We see, moreover, that the one really essential, universal, physiological property common to all living things is GROWTH. Crystals also grow, and in ultimate analysis it appears that there is essential community of nature between these two kinds of growth.

The tendency of modern science is to break down the partition between the organic and the inorganic; and to show that both are governed by the operation of forces which are the same in kind, though differently compounded.

Considered therefore in its widest sense, growth must be regarded as a necessary concomitant of evolution; and if evolution is universal, so is growth.

In each case growth is due to the universal tendency of like units to unite and of unlike units to separate. On this subject Mr. Herbert Spencer has expressed himself very clearly. He says:—"The deposit of a crystal from a solution is a differentiation of the previously mixed atoms, and an integration of one class of atoms into a solid body, and the other class into a liquid solvent. Is not the growth of an organism a substantially similar process? Around a plant there exist certain elements that are like the elements which form its substance, and its increase of size is effected by continually integrating these surrounding like elements with itself. Nor does the animal fundamentally differ in this respect from the plant or crystal. Its food is a portion of the environing matter, that contains some compound atoms like some of the compound atoms constituting its tissues; and either through simple imbibition or through digestion, the animal eventually integrates with itself, units like those of which it is built up, and leaves behind the unlike units."

Growth therefore is an integration with the organism of such environing materials as are of like nature with it. When the quantity of matter thus integrated exceeds that lost by disintegration, there is increase of size. It is to be regretted that modern physiologists have failed to give this subject that attention which its intrinsic importance demands.

We know, however, that all living things grow by taking into their substance new particles, which they dissolve and convert into fresh protoplasm. This they effect by intercalation of the new particles between the existing molecules, by a process of *intus-susception*. On the other hand, crystals grow by the addition of new particles to their exteior, by *accretion*.

These different modes of growth are consequent on the different states of density or molecular aggregation *inherent* to organisms and crystals respectively.

The semi-fluid or viscid state of aggregation is inherent to the former; the solid state to the latter. But in both cases the result is dependent upon the working of the same chemical and physical forces, of which the sun is the ultimate source. The molecular forces determine the form which solar energy will assume. Thus "one chain of causation connects the nebulous original of suns and planetary systems with the protoplasmic foundation of life and organisation."

In proceeding with our investigation into the

phenomena of growth in living things, vegetable life cannot be considered apart from animal life, since the same fundamental properties are common to both. Every one acquainted with Schwann's able thesis on the identity of the structure of plants and animals, will, I think, agree with me in maintaining, that an acquaintance with the chief facts of vegetable as well as of animal morphology, is essential for a scientific pathologist.

How otherwise can he take that comprehensive survey of the whole domain of vital phenomena which is so necessary? Moreover, in plants, the study of pathological processes is often much simpler than in animals, owing to the absence in them of many factors, such as the nerves and blood-vessels, which, in animals, often complicate and obscure the essential nature of such processes.

The material basis of life in both cases is the same —to wit, the nitrogenous carbon compound called protoplasm. The manifestation in this substance of those phenomena of motion, which we call *life*, is chiefly the result of two causes. In the first place of its characteristic semi-fluid state of aggregation, owing to the peculiar way in which a certain quantity of water is combined with its solid matter, whereby its units are so far freed from mutual restraints, that incident forces can readily effect changes in their relative positions. In the second place, to its extreme instability, owing to the easy decomposibility of its exceedingly composite albuminious combinations of carbon. It is chiefly by virtue of the latter quality that organic matter is so exceedingly sensitive to the influence of surrounding agencies.

In consequence of this extreme instability of its compounds, minute disturbances can initiate in it great reactions—setting up extensive structural modifications and liberating large quantities of motion.

In short, the molecules of this substance are of such kinds and so conditioned as easily to admit of re-arrangement.

Thus the constitution of protoplasm specially adapts it to receive and produce the internal changes, required to balance the external changes, the continuous adjustment of which, according to Spencer, constitutes life.

Before proceeding to treat of the various ways in which protoplasm is modified, we must first of all consider the manner in which its integral parts are united together; for whatever has come into being can only be known from the process by which it came into being.

I conceive that the simplest primitive living things must have been exceedingly minute, unstable, structureless molecules of protoplasm, nearly akin to those molecules of nitrogenous colloidal matter, into which all organic matter is resolvable, and like them possessed of varied atomic constitution.

It appears certain that the simplest living things now *extant*, such as the *Monera*, must have originated from these primitive elementary units, by a process of growth and modification in the gradual course of organic evolution.

These creatures are really nothing but small globular bits of fluctuating protoplasm. Though under all but the very highest magnifying powers they appear to have a perfectly homogeneous structure ; yet we know by their re-actions that they must really be possessed of a highly complex molecular structure, far beyond the range of vision. It even seems probable that they contain many different kinds of molecules already formed and individualised. In this case their various degrees of segregation may be regarded as the outcome of differentiations already established though invisible.

This doctrine of molecules bring the ultimate constituents of living things to one of the oldest and most universal ideas extant.* In his provisional hypothesis Pangenesis, Darwin suggests that each cell of the

* The five following propositions enunciated by Democritus fairly represent the *present* state of this theory. Strange that the progress of the ages should be so !

1. From nothing comes nothing. Nothing that exists can be destroyed. All changes are due to the combination and separation of molecules.

2. Nothing happens by chance; every occurrence has its cause, from which it follows of necessity.

3. The only existing things are the atoms and empty space ; all else is mere opinion.

multicellular organisms contains molecules derived from every other cell of the whole organism; and that each cell is constantly giving off such molecules, as minute gemmules, during the whole cycle of its existence.

Careful study of the phenomena of imbibition of water by protoplasm first led Nägeli to the hypothesis that the water penetrates between its molecules, each being surrounded by layers of varying thickness. He maintains that striation of the cell wall, &c., depends entirely on the quantity of water surrounding its particles. By examination with polarised light, he concludes that these ultimate particles are crystalline; but with this conclusion, Strasburger dissents. According to the latter observer, the molecules of protoplasm consist of multivalent atoms of solid substance, united by chemical affiinity; and the water present is retained in the intermolecular meshes by capillarity.

The *Monera* may be regarded as typical of organic units of form of the lowest grade—plastids or aggregates of the first order.

---

4. The atoms are infinite in number and infinitely various in form; they strike together, and the lateral motions and whirlings which thus arise are the beginnings of worlds.

5. The varieties of all things depend upon the varieties of their atoms, in number, size, and aggregation.

Atoms must be regarded merely as centres of force. From the simple assumption of such centres having attractive and repulsive forces all the general properties of matter—whether organic or inorganic—may be explained.

We may, I think, safely assume that such forms as the *Monera* have acquired their actual size and relative complexity of molecular structure, as well as their other essential qualities, in accordance with the laws of *Adaptation* and *Heredity*. Of these it is necessary to say just a few words.

The laws of *Adaptation* depend on the endless variability of protoplasm, owing to the unequal incidence of ever changing external conditions; which may operate either directly, or indirectly by survival of the fittest through natural selection.

The laws of *Heredity* depend upon the persistence of impressions (unconscious memory) in protoplasm. Hence every living thing produces new ones, *each after its kind*. It is by virtue of this property that—in the words of Sir James Paget—" a mark once made on a particle of blood or tissue, is not for years effaced from its successors."

Of all the properties of protoplasm this is the most characteristic, and it serves better than any other to mark off the organic from the inorganic.

When crystals are dissolved and recrystallised the same forms are reproduced again and again; but the combinations which determine organic forms repeat themselves hereditarily, and at the same time they undergo change.

These two sets of phenomena are so completely interwoven, that it is generally impossible to say how much of a given morphological change is attributable to the one, and how much to the other. Taken as a whole they stand in a certain opposition. Heredity is the cause of the stability of organisms, and Adaptation of their modification.

Adaptation is merely the material expression of change of function, but modification of function and its expression by obvious morphological change is a gradual process; hence adaptation is only noticeable in a long series of generations, whilst heredity can be recognised in every generation. The inherent tendencies of organisms are then proclivities *inherited* by them from ancestral organisms, which the past succession of living things has bequeathed. Thus *each cell's life the outcome of its former living is.*

All animal and vegetable organisms *commence* their existence in this simplest form, which is permanently retained by the lowest; but the permanent forms of the higher organisms result from the aggregation of these units variously modified. In the words of Oken, "every living thing has arisen out of slime, and is nothing but slime in different forms." Out of this substance the different organic structures may be moulded, either directly or, as usually happens, with various degrees of indirectness. The hypothesis of evolution compels us to this conclusion, which the facts oblige us to make. On this subject Spencer remarks "As structureless portions of protoplasm must have preceded cells in the process of general evolution, so, in the special evolution of each higher organism, there will be an habitual production of cells out of structureless blastema."

In the case of crystal formation we see that the whole aggregate exerts a force which constrains the newly integrated units to take up a certain definite form. The re-arrangements of organic units are determined in a precisely similar way. This hypothetical property, as to the real nature of which we are ignorant in both cases, is called *polarity*. Thus regarded, polarity may be defined as the resultant of the physico-chemical forces which determine molecular arrangement. In the words of the far-seeing Schwann, "Living things are nothing but the forms under which substances *capable of imbibition* crystallise." Hence the polarity of protoplasm is an important element in the evolution of organic forms.

In extending this analogy to the multicellular aggregates which comprise the great bulk of plants and animals, the *difference* between living things and ordinary crystallisation is brought out, owing to the obvious co-existence in their evolution of so many complex, varied and mixed conditions. Here we have a number of cells combined together within certain limits—forming what has aptly been called a society—which do not enter into the combination as dead particles; but each undergoes the particular series of changes peculiar to its own developmental cycle—produces new ones which repeat the same series of changes, each after its kind—then dies, and is cast out or absorbed. Thus the arrangement and form of the whole organism is constantly changing in all its parts; there is nothing definitely fixed about it, every particle is in a constant state of fluctuation.

In the ordinary course of nature all living matter proceeds from pre-existing living matter, a portion of the latter being detached and acquiring a more or less independent existence.

This is reproduction or growth of the organism beyond its individual limit of size. It is essentially *cell-multiplication or discontinuous* growth, so called in contradistinction to that mere increase of individual size which is *continuous* growth. When a cell divides, the resulting cells are at first smaller than the original cell. Under continuous growth are included all the phenomena concerned in determining the composition and size of the organism, whereby the balance between the processes of waste and repair is maintained. Its chief abnormalities may be comprised under the heads of Hypertrophy, Atrophy and Degenerative Metamorphoses.

In every act of reproduction a certain quantity of protoplasm is transferred from the producing to the produced organism, and along with it is the molecular motion peculiar to the parental individual. The phenomena of heredity are essentially dependent upon this material continuity and partial identity of the producing and produced organisms. If, instead of a succession of individuals thus produced, we substitute in imagination, a single continuously existing individual, we shall at once see the relation in which an organism stands to the rest of its species, and how the succession of all living things is, as Goëthe says, a linked chain.

In the lowest forms of life the various functions are at first exercised equally, or nearly so, by all parts of the protoplasmic body. Yet even these undergo, concurrently with growth, a series of chemico-physical changes from the beginning to the end of their existence, by which their originally homogeneous substance is variously metamorphosed, in accordance with the degree of specialisation attained in the physiological division of labour. Thus a nucleus is differentiated from the homogeneous protoplasm, and frequently a limiting membrane, &c.

It is now customary to speak of both protoplasm and nucleus as the essential constituents of a typical cell (fig. 1.) Until quite recently, however, the latter was regarded as only of secondary importance. The researches of Strasburger, Bütschli, Flemming, and others have now demonstrated the great importance of the nucleus in the reproductive process.

The cell *protoplasm* is found to consist, under the

highest powers of the microscope, of a fine fibrillar network, the meshes of which contain a homogeneous fluid. It is often thickly beset with small granules (microstomes) which appear, when highly magnified, as minute dots, and are probably finely divided nutritive matter. In all but the youngest cells a cell-wall is differentiated from the protoplasm.

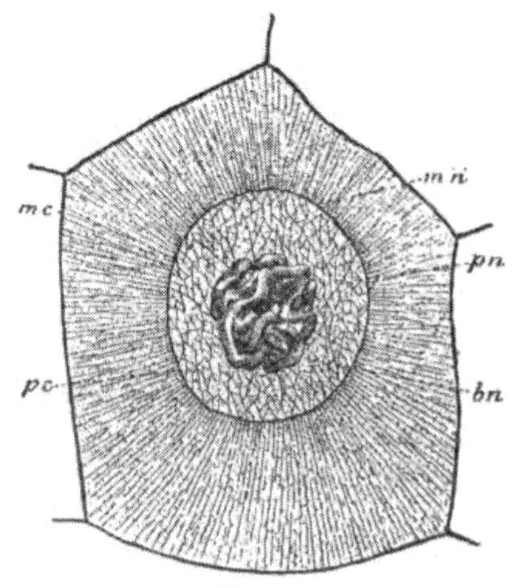

FIG. 1.

A typical epithelial cell. *m c*, cell wall. *p c*, cell protoplasm, showing radiating *reticulum*, enclosing fluid *plasma*. *m n*, wall of nucleus. *p n*, nuclear substance, showing *reticulum* and *plasma*. *b n*, contorted band of *nuclein* or *chromatin*.

The *nucleus* usually presents as a more or less globular vesicle embedded in the protoplasm. It consists of a very fine protoplasmic reticulum, the meshes of which are filled with a granular fluid. In addition to these achromatic substances the nucleus possesses, according to Carnoy, a distinctive substance called *nuclein*, or from its staining readily, *chromatin*. In young cells this substance exists as a long irregularly contorted filament (fig. 1); but in more mature cells it becomes condensed into one or several spherical masses, which constitute the *nucleoli*.

Embryonic cells and all cells in active growth have very large nuclei, with but a narrow zone of surrounding protoplasm.

The typical cell form is more or less spherical—growth being equal in all directions. Modified cell shapes are produced by inequalities of growth. Rapidly growing cells are in a state of tension from the imbibition of water which causes them to swell up. Being also extensible and flexible, growth ensues in the direction of least resistance, which is determined by the mutual pressure of adjacent parts. Every cell after its formation begins at first to grow slowly, then reaches its greatest rapidity of growth, and ultimately growth declines, until at last it ceases.

Thus the form and substance of the cells are altered, and at the same time their physiological properties are altered, so that new relations are established.

Specialization of the cells is the cause of the great variety in organization, which is much more marked in the animal than in the vegetable kingdom. In the multicellular aggregates, most of the originally like units (bastomeres) undergo analogous changes, which result in the differentiation of the various tissues and organs. All such changes are included under the term development, which is often held to be fundamentally distinct from growth. However this may be, it is certainly convenient to bear the distinction in mind. In all cases development is a change from the general to the special. Every living thing, arising out of apparently uniform matter, advances to multiformity through insensible changes. In the words of Harvey: "there is successive differentiation of a relatively homogeneous rudiment into parts and structures which are characteristic of the adult." The production of all organic forms is in fact by the slow accumulation of modification upon modifications, and by the slow divergencies resulting from the continual addition of differences to differences. In this respect the development of the aggregate as a whole, repeats the development of its component units.

Thus each simplest living thing has a certain natural period of existence in the ordinary conditions of active life; at the end of this period, if not previously destroyed by some outward force, it degenerates and dies, and its substance is resolved into more highly oxidised compounds of its elements. Such is the cyclical series of changes through which every living thing must go. Each cell of a multicellular organism, when once formed, leads normally a double life; it continues to grow of its own inherent power, to the attainment of its own special development; and at the same time it is directed by the influence of the entire organism, in accordance with the requirements of the specific hereditary tendency of the whole. In these organisms therefore disease may be due either to abnormal states of the cells, or to disturbances in their co-ordinating mechanism.

All sorts and conditions of growth in the end will be found to depend upon the molecular processes of *Nutrition*—understanding by this term the whole of the material changes wrought in the organism through the influence of the surrounding outer world. These many complicated conditions are never absolutely identical for any two individuals, hence endless variability is a universal property of all organisms. Change of nutrition is unquestionably the true cause of all morphological variations. Herein lies the possibility of evolution, which must in all cases be understood to be the direct or indirect outcome of the unequal incidence of these ever-changing conditions. If it were possible to expose all the individuals of a species during many generations to absolutely uniform conditions of life, there would be no variability.

Growth being an integration with the organism of such environing matters as are of like nature with it, is necessarily dependent upon the available supply of such materials; of these the most essential is water, for it is certain that living things cannot grow without it, however abundant the other requisites may be. Light, heat, and nutriment are also important factors. The quality and quantity of nutritive material at the disposal of the cells have a profound influence upon their behaviour. It depends upon the nutriment supplied to the female larva of a bee whether it will become a neuter or a sexually perfect female; and the sexual perfection of many internal parasites is similarly dependent upon the food supply. In reproduction or cell-multiplication, to which all organic units are inherently prone, the occurrence and nature of the process, whether by agamogenesis, gamogenesis, or alteration of generations, is determined by the conditions of nutrition. Impregnation must be regarded merely as one of the many conditions which affect the process. The influence of the male element on the germ-cell may, as Caspar Fried Wolff suggests, be compared to that of a kind of nutriment.

Blood and lymph vessels and nerves have not the slightest *direct* influence on the processes of growth. These are only of importance in so far as they regulate the supply of nutriment, &c. In plants, the early embryo, and the lowest animals, all the phenomena of growth go on without them.

The growth of every living thing is thus determined in accordance with one common formative principle, which manifests itself under many different aspects, but as to the real nature of which we are ignorant.

On this subject Paget remarks: "Of this force, by whatever name we designate it, whether as the formative, or the plastic, or, more explicitly, as the force by which organic matter, in appropriate conditions is shaped and arranged into organic structure; of this force and of those that co-operate with it, we can, I think, only apprehend that they are, in the completed organism, the same with those which actuated the formation of the original tissues in the development of the germ and of the embryo. As we have seen that the new formation of elemental structures in the maintenance of tissues is a repetition of the process observed in their first development, so we may assume that the forces operative are the same in both processes."

Organic growth has its limits—why is this? Spencer has answered the question as follows: growth varies according to the surplus of nutrition over expenditure, and is unlimited or has a definite limit, according as the surplus does or does not progressively decrease.

Why should growth tend in the direction of cell multiplication, rather than to the production of large unicellular aggregates? We must seek the answer to this question in the conditions of molecular cohesion in protoplasm. In all cases the process is evidently one of disintegration, and as such, opposed to that integration which constitutes the individual. Diminished nutrition after the organism has attained a certain size, leading to lessened growth, is, according to C. F. Wolff, the chief cause. For cell multiplication does not take place whilst the parental individuals are growing rapidly; that is, whilst the process of growth greatly exceeds the opposing forces, but it begins when nutrition is nearly equalled by expenditure.

CHAPTER II.

REPRODUCTION.

"If ever we are to escape from the obscurities and uncertainties of our art, it must be through the study of those highest laws of our science, which are expressed in the simplest terms in the lives of the lowest orders of creation. It was in the search after the mysteries—that is, after the unknown laws—of generation, that the first glance was gained of the largest truth of physiology; the truth of the development of ova through partition and multiplication of the embryo cells."

PAGET.

I SUPPOSE it must be in consequence of some deeply ingrained vague instinct of the truth, that the human mind in all ages, climes and zones, manifests such irresistible tendency to penetrate the mystery of the beginning of things, and above all the origin of living things, including our own origin.

The quotation at the head of the chapter admirably expresses this feeling; its winged words fly as it were by inspiration of genius, to the heart of the matter. Thus it is, and not by method alone, that now and again, as the days of the years roll by, men arise to lighten the darkness which surrounds all knowledge.

It is only in comparatively recent times that the view has been generally adopted, that diseased states are merely modifications of healthy states—perturbations of the normal life, caused by changed and abnormal conditions of existence. By health we merely mean that the vital functions are being performed in a manner which experience has led us to regard as normal. Similarly by disease we merely imply a phase of life whose manifestations deviate in some way from the normal. These terms are but relative conventional conceptions.

We infer disease from the appearance of some abnormality among the accustomed manifestations of life. Thus at the outset our conception of disease is a purely physiological conception. There is no essential difference between the laws which govern physiological and pathological processes. The difference lies in the conditions under which the organic forces and substances operate. It is necessary to insist upon these elementary notions, because even now the full significance of the great principle of modern pathology—that every pathological process has its physiological prototype—is far from being generally recognised.

It is impossible to say in any given case where health ends and disease begins. Thus, no one can determine where what we call normal morphological variations end, and cancer and tumour formations begin. All that we can say is, that when structural or functional changes are hurtful, they belong to the province of pathology. The study of healthy processes must therefore necessarily precede that of the phenomena of disease. "There can be no question," as Mr. Huxley remarks, "as to the nature or value of the connection between medicine and the biological sciences. There can be no doubt that the future of pathology and of therapeutics, and therefore of practical medicine, depends upon the extent to which those who occupy themselves with these subjects are trained in the methods and impregnated with the fundamental truths of biology."

In the lowest organisms the only obvious phenomena of disease, are such abnormal states of their protoplasm as I have mentioned in the previous chapter. But in the case of the higher organisms—where each of the so-called individuals is really a kind of society, in which a number of mutually dependent cells are bound together, so that whilst each is dependent upon others, yet each has its own special action by which it effects the performance of its own duties—disease may consist either in abnormal states of the protoplasm of the component cells, or in disturbances of their co-ordinating or alimentary machinery or in both. In consequence of such disordered conditions cell-multiplication may arise at a *place* where it has no business, or at a *time* when it ought not to occur, or to an *extent* which is at variance with the normal formation of the body.

It is to the study of such conditions as these that we must now direct our attention.

The first subject of which I propose to treat, in this connection, is that of generation or reproductive growth in its various aspects. In approaching this subject we must first of all divest ourselves of the ordinary idea that the union of the sexes is the most important and essential thing in the process of reproduction. As I have previously stated, the essential thing in reproduction is growth of the organism beyond its individual limit of size, or *cell multiplication*. Impregnation must be regarded merely as one of the many conditions which determine or affect the process. In all cases reproduction is intimately dependent upon the molecular processes of nutrition.

This simplest mode of reproduction, by cell multiplication, is really the most important and widely spread of all such modes, for in ultimate analysis they can all be reduced to it.

I propose to show that there is no fundamental distinction between the various modes of reproduction—sexual and a-sexual, the process of repair and formation of tissues, the reproduction of lost parts, and the various morphological variations, including bud, cancer, and tumour formation.

I maintain that all of these apparently so different

phenomena, are merely modifications of one common process, which underlies and is the cause of them all —to wit, cell growth and multiplication.

Further, I shall endeavour to show that in all cases the *real meaning* of cell multiplication is the *production of new individuals.*

To this end I shall adduce facts showing that there is no warrant for the belief so generally entertained, that sperm cells and germ cells are fundamentally different from one another, and from all other cells. On the contrary I shall show that they differ from other cells only in this—that they are comparatively *less specialised.*

In the next place I shall give reasons for believing that every component cell of the multicellular aggregates has the inherent power, under favourable conditions, of developing itself into the form of the parent organism; in short, that each cell is *potentially* the whole organism.

That the degree of development *actually* attained by each cell falls far short of this in most cases, is due to the restraining and modifying influence exerted by the cells of the whole organism on its protoplasm, which is thus compelled to the performance of comparatively subordinate, modified functions. Under these circumstances the primitive force remains latent, but it may often be evolved in varying degrees under favourable conditions, and sometimes even to its full extent. Thus there are plants in which any single cell separated from the rest suffices for the reproduction of the whole organism, and among animals nearly similar conditions prevail in the *Hydra*. These cases will be more fully described further on.

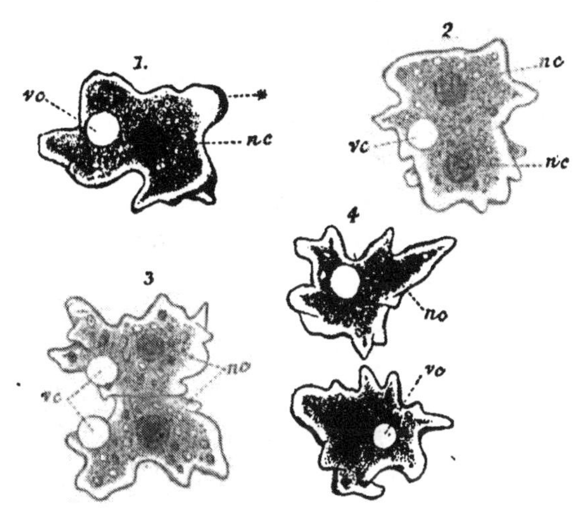

FIG. 2.

Successive stages in the life history of an Amœboid organism, kept under constant observation for three days.—Zeiss D. 3.

1—The locomotor phase ; 2, 3, 4—The reproductive phase in successive stages ; 2—Nuclear division ; 3, 4—Cell division.

*—Protrusion to form a pseudopodium.

nc—Nucleus.

ve—Contractile vacuole.

Hence I have arrived at the important conclusion that the processes of repair and reproduction of lost parts, and the various morphological variations, including bud, cancer and tumour formations, are nothing but more or less abortive attempts of certain cells to reproduce new individuals. Whence it follows that the laws of reproduction are also the laws of cancer and tumour formation. I state this most emphatically, because the doctrine has never before been clearly enunciated, although it has for some time been vaguely foreshadowed. The study of reproduction and its laws is, therefore, an essential preliminary to the study of the ætiology of Cancer and Tumour formation. It is the physiological prototype of these pathological processes.

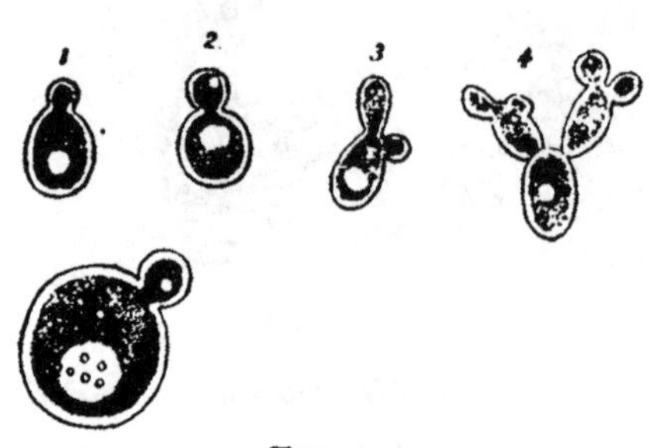

FIG. 3.

Nos. 1 to 4 cells of ordinary brewer's yeast (torula cerevisiæ), showing stages of bud formation, leading to the formation of branching colonies. —Zeiss D. 4.

No. 1 of the above cells as seen under 1-16th immersion.

The simplest living things, such as the *Monera*, reproduce themselves by *Fission* or simple division (fig. 2).

Gemmation, or bud formation, is only a modification of this process (fig. 3).

It appears certain that in the earliest period of the organic history of the world, all organisms propagated themselves in this way, which must therefore be regarded as the original primitive method.

When a cell contains a nucleus, its division usually precedes that of the surrounding protoplasm. In contradistinction to the above process, this is called indirect cell-multiplication or karyokinesis.

At its commencement the nuclear membrane and the nucleoli disappear, and the nucleus increases in size. Its chromatin assumes successively the forms of a coil, a wreath, and a star (fig. 4).

By this time its achromatin has been transformed into a nuclear spindle, and the chromatin star lies at right angles to its equatorial plane.

This star next divides longitudinally into two equatorial groups, each of which travels towards an opposite pole of the spindle.

Each of the groups then becomes converted into a star—thus we have the di-aster stage. These stars are the rudiments of the daughter nuclei. Their separation is completed by rupture of the remaining filaments of the spindle. By passing through the converse series of changes, the daughter stars are gradually transformed into new nuclei. During the di-aster stage active movements take place in the cell protoplasm, which becomes constricted, and finally completely divided, by the time the newly-formed nuclei have reached the coil stage. The amount of

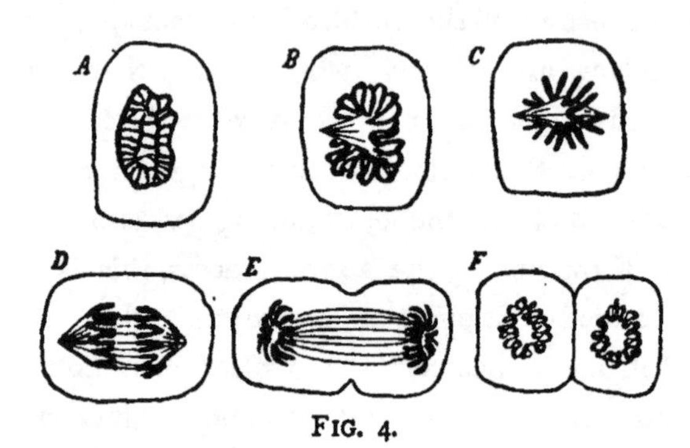

FIG. 4.

A-F.—Karyokinesis of a tissue-cell. *A.* Nucleus preparing for division—coil stage. *B.* Wreath stage—the chromatin arranged in complicated loop around the equator of the achromatin spindle. *C.* Aster stage. *D.* Division of the aster into two halves, each of which migrates to the opposite pole of the spindle. *E.* Di-aster stage—the chromatin forms a star at each pole of the spindle. *F.* Daughter cells in wreath stage—the newly-formed nuclei passing through retrogressive changes to reach the resting stage.

protoplasm which passes over to each nucleus may be equal (Fission, fig. 4), or one may take much more than the other (Gemmation, fig. 5). During karyokinesis a radial arrangement is often noticeable in the cell protoplasm surrounding each daughter nucleus.

This is the typical process, but variations occur

in the different classes of plants and animals. All animal and vegetable cells, whether of normal or pathological origin, arise in this way. In the protozoa the direct method prevails, but in the metazoa it is quite exceptional. When this occurs it may be regarded as due to reversion to the ancestral method.

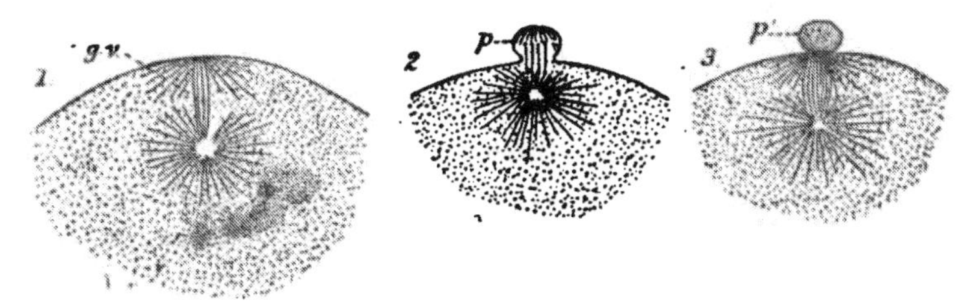

FIG. 5.

Stages in the formation of the first polar cell, by budding, in the ovum of a star-fish.

gv—Germinal vesicle undergoing division transformed into a spindle-shaped system of fibres (karyokinesis).

p—The polar cell becoming extruded from the surface of the ovum.

p'—The polar cell completely extruded.

In the lowest organisms, these processes result in the formation of *new individuals* by direct transformation of the germs, as the detached portions of protoplasm are called. In the case of fission, the growth which causes the process is total, affecting the whole protoplasm (figs. 2 and 4), so that in most cases neither of the resulting cells can be called the parent of the other. But in gemmation, a portion of protoplasm, already more or less developed, separates from the producing individual to form a new being (figs. 3 and 5).

Here the resulting individuals are of unequal value; the one is the parent of the other. The growth that causes the process is partial, affecting only a portion of the parental organism. Hence the whole difference between fission and gemmation is quantitative merely, and consists in the different amounts of protoplasm, given over by the producing to the produced organisms. In like manner each of the constituent cells of the higher organisms reproduces its kind, whereby the formation and maintenance of their various tissues and organs is ensured.

Endogenous cell formation, the so-called free cell formation and the like, which were formerly described as essentially different processes to the above, have recently been shown to be but modifications of it. Fission and gemmation when incomplete lead to the formation of colonies. When the nucleus divides and the protoplasm grows without being marked off into corresponding separate portions, large multinucleated masses of protoplasm—the so-called giant cells—are formed. A similar result ensues when, as sometimes happens, several separate cells fuse into one continuous mass. Such structures are found in many normal and pathological tissues.

The subject of reproduction by fission and gemmation in the case of the multicellular organisms, will be discussed in a future chapter, when treating of bud formation in general. It will then be shown that each new individual thus separated is the product of the multiplication of but a single parental cell. Between these modes of reproduction and that by "germ buds" there is no real distinction.

In these cases the new individuals thus arising do not separate from the parental form until they have undergone considerable or even complete development.

In the case of *spore formation* a *single* cell separates spontaneously from the surrounding cells of the parental organism, and having left this, multiplies by division, and so develops into a new many-celled individual, as among many of the *cryptogamia*, &c.

This mode of reproduction differs from the foregoing only in that separation of the reproductive cell from the parent takes place before it has become subject to the process of cell multiplication.

In all these cases the development of new individuals from the germ cell takes place independently of other living matter. This is *agamogenesis* or a-sexual reproduction.

Its highest development, spore formation, leads directly to a form of reproduction in which the germ cell cannot develop into a new individual of itself; but must first be fructified by contact with living matter *different* from it—the sperm cell. This is *gamogenesis*, or sexual reproduction.

In the lowest organisms gamogenesis has never been observed, and in the highest organisms agamogenesis is absent. It appears to be a general rule throughout both the animal and vegetable kingdoms, that the product of gamogenesis is more highly organised than that of agamogenesis. Hence we are led to infer that a more complex kind of development is initiated by sexual reproduction.

Careful examination of the subject in its different bearings shows that there is no fundamental difference between gamogenesis and agamogenesis, and that the former has originated from the latter at a comparatively late period of the earth's history, in the first instance apparently through spore formation.

The two different generative substances essential for gamogenesis may be produced either by the same individual (monœcious or hermaphrodite), or by two different individuals (diœcious).

The most ancient primitive form of this process appears to have been through double-sexed individuals, as it at present occurs in the majority of plants and in a few animals, such as garden snails and many worms. Some hermaphrodites can fructify themselves, but in most cases copulation, or the reciprocal fructifying of two hermaphrodite individuals is necessary for reproduction. Here we have evident indication of transition to sexual separation, which is universal amongst the higher animals, but occurs only in a few plants, *e.g.*, vallisneria, willows, poplars, &c.

Thus the distinctive feature of gamogenesis under all its different aspects is the coalescence of a detached portion of one organism, with a more or less detached portion of another allied organism, or of two allied unicellular organisms. In a few instances these detached portions are derived from different parts of the same organism.

In its simplest condition—known as *conjugation*—these two factors are apparently similar; but there is usually well marked morphological difference between them. The relatively large and passive factor is the female element, the *germ cell;* the relatively small and active factor is the male element, the *sperm cell.* Throughout the whole organic world the germ cell almost invariably maintains the form of a spheroid of protoplasm; but the form of the sperm cell is subject to great modifications, especially in the direction of the development of flagellate processes.

Fructification or impregnation is the result of the union of these two elements; the essential nature of which is the physical admixture of protoplasmic matter derived from two *different* sources. The simple cell—*cytula*—resulting from the fusion, develops into the new being. This cell must be regarded as an entirely new independent formation, since in it the characters of *both* parents are potential.

Such is the starting point which in the higher plants and animals initiates the evolution of every new individual. Viewed in this light gamogenesis is seen to be nothing but a particular case of cell multiplication, and impregnation merely one of the many conditions which affect the process. As I have previously mentioned, the influence of the sperm cell on the germ cell may be compared to that of a kind of nutriment; but with this proviso, that in this nutriment, no less than in the germ cell, the specific form of the parental organism is involved with all its individual peculiarities.

Thus in every kind of reproduction the essential thing invariably is the detachment of a portion of the parental organism, possessing the capability of more or less independent growth. Such detached portions may consist either of a single cell, or of a group of cells derived from a single cell, variously developed.

Very interesting are the phenomena of *Parthenogenesis* or virginal reproduction, as representing a transition from sexual to a-sexual reproduction. Here germ cells, which often appear to be formed exactly like ova, develop themselves into new individuals, without the influence of any fructifying sperm. It occurs in certain plants and animals, and is probably of the nature of a relapse from the sexual method.

Among our common honey bees a male individual (drone) arises out of the egg of the queen, if it has not been fertilised; a female (queen or working bee) if the egg has been fertilised. This shows there is no such fundamental distinction between gamogenesis and agamogenesis as is commonly supposed.

In true parthenogenesis there occurs, along with gamogenesis, in a true ovarium or homologous organ, a form of agamogenesis, exactly like gamogenesis, save in the absence of fertilisation, as in silkworm moths. False parthenogenesis occurs when new individuals arise from *buds* in pseud-ovaria, which are not ova properly so called as in *aphides*.

This process is intermediate between true parthenogenesis and that form of agamogenesis called by Owen *metagenesis*, in which new individuals *bud out*, not from any specialised organs, but from *unspecialised* parts of the parent, which are generally external, but may be internal, as in *distoma*.

We now pass to the very remarkable series of events known as the "*alternation of generations*." Until comparatively recently it was believed that in every species the successive generations were always alike—*homogenesis;* but it is now known that this is not always the case—*heterogenesis*.

Many plants and animals produce a generation unlike their parents; these may produce others like themselves, or like their parents, or like neither and so on; but eventually the original form reappears. Here gamogenesis alternates with agamogenesis or parthenogenesis.

Since all gradations between complete alternation of generations and simple agamic budding combined with gamogenesis can be traced in forms now extant, it appears certain that the ancestors of the forms which now exhibit alternation of generations, originally reproduced themselves at the same time sexually and a-sexually, though probably the two modes of reproduction did not take place at the same season.

This subject is so important for the elucidation of my argument, that I must now give two examples of it, in some detail. For this purpose I have in the case of animals, selected the plant lice (*aphides*), because in these insects the whole process has been admirably worked out by Owen, Huxley and others, of whose works I have freely availed myself.

The impregnated ova of the *aphis* are deposited at the close of summer in the axils of the leaves of the plant infested, retaining their vitality throughout the winter; these ova are hatched by the returning warmth of spring: a wingless hexapod larva is the result. In the pseud-ovaria of these imperfect females, there bud forth pseud-ova which rapidly develop into similar imperfect females. At this season no winged males have appeared. This process of agamic multiplication continues throughout the summer. If the external conditions, such as warmth and nutriment, continue favourable, eight or more successive generations may be thus produced. But when the weather becomes cold and the supply of sap fails in the plant, perfect males and females are produced, which by gamogenesis reproduce fertilised ova, thus completing the cycle.

Further experiments have shown that in such cases the rapidity of the agamogenesis is proportionate to the degree of warmth and nutrition, and that if the temperature and food supply be artificially maintained, the agamogenesis continues throughout the winter. When the favourable conditions have been kept up for several successive years, agamogenesis has likewise continued. In short, it seems probable that this agamic reproduction could be continued indefinitely, if all the requisite conditions are fulfilled.

This connection between reproduction and such diminished nutrition as makes growth relatively slow, was first fully made known by the celebrated German biologist Caspar Friedrich Wolff, in his *Theoria generationis.* Wolff first shows that the organs of fructification in plants are merely modified leaves. The plants themselves he regards as branched colonies of individuals, of which the leaf-bearing axes are the agamogenetic and the flower-bearing axes the gamogenetic individuals. He next demonstrates that the modifications which leaves undergo in the formation of flowers, are the result of diminished nutrition, leading to imperfect growth. Hence at first, when nutrition is at its best, only leaf-bearing axes or asexual individuals are produced. But as we approach the flowering axes, the influence of diminished nutrition shows itself in the production of smaller leaves and shorter internodes. Finally, in the flower itself, the non-development of the terminal internodes, the very stunted modified leaves from which the organs of fructification proceed, and the situation of the whole at the extremity of the axis—far removed from the main channels which supply nutriment—all indicate that extreme diminution of nutrition and approximate arrest of growth which characterises the advent of gamogenesis. In short, the entire structure of the flower suggests that of an abortive sexless axis.

Further, if a plant known to put forth leaves under ordinary circumstances a certain number of times, for instance six times, before developing flowers, is set in very poor soil, not only will its leaves become very small and imperfect, but it will scarcely have put forth leaves three times before fructification occurs. Whereas if this same plant is placed in very moist and rich soil, its leaves will become larger and more perfectly formed, and instead of developing leaves six times, it will put these forth nine times or more before organs of fructification appear.

Moreover, if whilst fructification is thus delayed by the richness of the soil, the plant is transplanted to poor soil, flowers immediately appear.

Lastly, if a plant which has already developed the calyx and rudiments of the corolla and anthers in poor soil, is quickly transplanted into rich soil, the anthers will be seen to undergo transformation into petals, in consequence of the excess of nutrition. When the petals and stamens are thus changed into green leaves the return from gamogenesis to agamogenesis is still more marked.

In accordance with this principle, trees may be made to fruit when quite young, by cutting their roots or confining them in small pots; and luxuriant branches that have had the flow of sap checked by "ringing," soon produce flowers instead of leaves.

Seeing that the action of the sperm cell on the germ cell was the cause of its development—for before it was deficient in this respect—and having in mind the above consideration, Wolff was led to regard the former as nutriment in its highest perfection, supplied to the germ cell from without, instead of through the ordinary channels.

Under what circumstances do single cells develop into new individuals? How have such cells acquired their wonderful properties? And how do they differ from other cells of the organism? These are some of the questions that naturally arise out of such considerations as the foregoing. In answering them it is necessary first of all to bear in mind the fact that simple undifferentiated protoplasm is the proximate source of all growth, reproductive and otherwise. Further, we must remember that every detached portion of such protoplasm has the inherent power of arranging itself into the form of the organism whence it was derived.

In the simplest living things, as I have previously mentioned, the various functions are exercised equally, or nearly so, by all parts of the protoplasmic body; hence when the mass divides each part retains all the functions of the whole, including the reproductive function, and so constitutes a germ whence the entire organism can be reproduced.

The question here presents itself,—how small a portion of the multicellular organisms may contain the force necessary for the reproduction of the species? In the highest organisms which propagate only by gamogenesis, this force is found only in the germ cells (ova), whilst all the other parts of their bodies are destitute of it. In those plants and animals which propagate by buds or sprouts, the germ consists of a mass of cells which may be produced at almost any part of the body of the parent. In some of the lower animals the same reproductive power is possessed by every aggregate of cells. In other lowly organised beings the power of reproducing new individuals is not merely manifested by separate portions of most parts of the body, but in some cases sub-division, even to the ultimate morphological units, does not destroy this power; isolated cells are, in fact, adequate to propagate the species.

When, therefore, in such organisms as the *Hydra* and many plants, we see protoplasmic cells—derived from many various parts in the same individual or in different individuals—separating themselves from the rest of the organism and independently developing into new individuals, we may, I think, fairly conclude that every like constituent portion of all organisms possesses similar power of independent growth and reproduction. Similarly, we may infer that each of the number of cells (blastomeres) into which, in the higher organisms, the germ cell at first divides, also retains all the properties of the germ itself, and is capable of reproducing the entire organism.

Now, if this be so, that every component cell of the multicellular organisms has this inherent power of reproducing the entire individual, how then does it happen that these cells usually remain aggregated together, and unite only in the form proper to the species?

The answer to this question is threefold.

In the first place, in order that this inherent power may manifest itself, the conditions must be favourable to its development. It cannot reasonably be doubted that each unit of an organism acts, by virtue of its polarity, on all the other units, and is reacted on by them. This normal restraining influence, exerted on each cell by all the other cells, in accordance with the specific hereditary tendency of the whole, must be weakened, modified, or withdrawn, before the potential reproductive power can become actual. In the ordinary course of normal life the only cells thus set free are the reproductive or germ cells. But *any* cells, abnormally emancipated from the controlling influence of the whole, may then grow and multiply independently, or with various degrees of dependence, according as the emancipation is complete or more or less incomplete. In lowly organised beings, such as the *Hydra*, this controlling influence is very feeble, so that almost any cell in these creatures may develop itself into a new individual. But in the case of the higher organisms this controlling influence of the whole is much more powerful, except in certain definite localities, hence the formation of reproductive cells is limited in them to special organs, and all their other cells are deficient in this respect.

The second factor in the explanation seems to be that in consequence of the special metamorphosis of their protoplasm, most of the cells of the higher organisms suffer loss or impairment of their reproductive power. In the course of development from originally like units (blastomeres), as the process of division and differentiation proceeds, many cells acquire special functions, and much of the original protoplasm is used up and converted into special tissues.

In proportion as the cells are thus specialised, we may infer they lose their primitive general functions. Thus their power of reproducing the whole organism may be greatly reduced or altogether lost, owing to all or the greater part of the original protoplasm being specialised and used up. But in the higher organisms certain cells never attain a high degree of development; they remain in a lowly organised condition and serve, according as they are more or less *unspecialised*, either as germs for reproducing the entire individual, or for forming and maintaining the various tissues and organs. Such cells are found in all growing parts; they are *the only real cancer and tumour germs*, as I shall hereafter show.

Hence most of the component cells of organisms produce only cells of the same kind as themselves, and are totally incapable of acting as germs of the entire organism, *e.g.*, epithelium cells always reproduce epithelium cells; hence also certain parts of the tentacles of hydras, when cut off, alone of all parts of the body fail to develop into new individuals.

In support of this view it may be mentioned that all kinds of reproductive cells and organs are composed of tissues remarkable for their low degree of organisation. In plants, as we have seen, they arise at the extremities of axes, where structure is least specialised. The male and female organs at first consist each of a solid bud of similar unspecialised, protoplasmic cells,

At this stage their likeness is so great that no distinction of sex can be made in the nascent flower; it is only subsequently that the difference becomes perceptible. The outermost cells then form the epidermis, and from hypodermal cells, immediately beneath this, the mother cells of the pollen and embryo sac arise. In the lower animals, such as the hydra, sperm cells and germ cells originate from epithelial cells that are but little differentiated. Similarly in the higher animals, the sexual elements arise from cells of the cœlomic epithelium, which have undergone very little differentiation.*

According to this view, then, the power of reproducing the whole organism will be limited to those cells which have remained unspecialised, and consequently have departed but little from the original

---

* "The relation of the reproductive elements to the primitive layers of the germ is as yet uncertain. E. van Beneden has brought forward very strong evidence to the effect that in the Hydractinia the spermatozoa are modified cells of the ectoderm, and the ova of those of the endoderm; but whether it can be safely concluded that this rule holds good for animals generally, is a question that can only be settled by much and difficult investigation. The fact that in the Vertebrata, the ova and spermatozoa are products of the epithelial lining of the peritoneal cavity, appears at first sight directly to negative any such generalisation. But it must be remembered that the origin of the mesoblast itself is as yet uncertain, and that it is quite possible that one portion of that layer may originate in the ectoderm, and another in the endoderm."—HUXLEY.

primitive type in which the organism commenced its existence.

The more widely diffused such cells are in any given case, the more readily will reproduction by budding or fission take place; the more localised such cells, the more limited the part in which this could occur. Thus, as Huxley remarks, "the gradual disappearance of agamogenesis in the higher animals would be related with that increasing specialisation of function which is their essential characteristic; and when it ceases to occur altogether, it may be supposed that no cells are left which retain unmodified the powers of the primitive embryo-cell."

In the third place, in order that this inherent power of the cells may manifest itself, the external conditions must be favourable to its development.

Granted all of these conditions, and even single epithelial cells from the leaves or stem, as in the case of *Begonia phyllomaniaca*, may become endowed with the power of reproducing the whole organism. Under similar circumstances in many plants and animals, a very small fragment of but little differentiated tissue may be capable of reproducing a whole organism, like that whence it was derived.

Now if, as I have maintained, sperm cells and germ cells are peculiar only in respect to their low grade of organisation, in which they both resemble unspecialised cells, it follows that they cannot be essentially unlike each other. Many facts point to this conclusion. The most striking are those recorded by Mr. Salter. He has described a monstrosity in two kinds of passion flower and a rose, in which ovules graduated into anthers, and produced pollen in their interior.

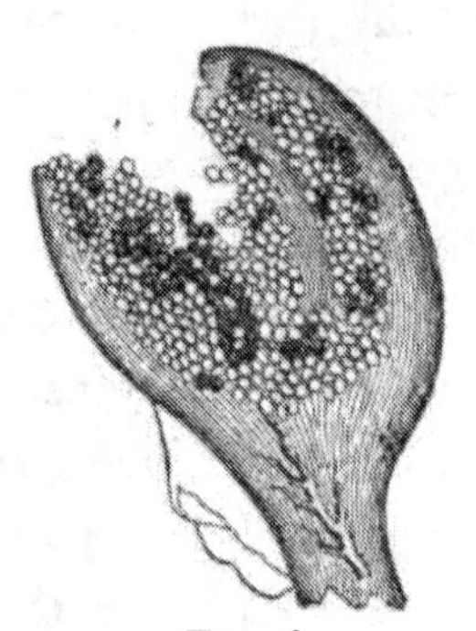

FIG. 6.
Formation of pollen within the ovule of *Passiflora.*

His description of one of these cases is as follows :—

"The pollen bearing ovules presented various intermediate conditions between anthers and ovules. Commencing at the distal extremity of the carpel was a bilobed anther, and passing in series to the base of the ovary, an antheroid body of ovule-like form, a modified ovule containing pollen, an ovule departing from a perfectly normal condition only in the development of a few grains of pollen in its nucleus, and finally a perfectly normal ovule." He subsequently makes this remark :—" For an ovule to develop pollen in its interior is equivalent to an ovum of an animal body being converted into a capsule of spermatozoa. It is a conversion of germ into sperm, the most complete violation of individuality and unity of sex."

As to the precise circumstances which determine the cells of the homogeneous rudiments of ovules, in one case to form mother cells and pollen grains, and in the other the embryo sac and germ cells, all that can be said is that position and external conditions appear to have some influence. For in all these cases of polleniferous ovules, the ovular structure had been exposed on an open carpel, instead of being confined within the ovary in the usual way.

In animals we may instance the hermaphrodite glands of Gasteropoda, where some of the similar epithelial cells develop into ova and others into spermatozoa.

Thus, then, there is no reason for believing that sperm cells and germ cells are fundamentally unlike one another and unlike other cells. They are simply unspecialised cells which have departed but little from the original general type.

This is borne out by the fact that, throughout the organic world, the less differentiated organisms are those most prone to agamogenesis, that is to say, their individual cells are possessed of a high degree of reproductive power; whereas in the more highly differentiated organisms, the absence of agamogenesis indicates great falling off in this respect.

In the highest organisms the reproductive cells, even when favourably situated with regard to nutrition, are of themselves incapable of further development. Though unspecialised they have lost the power of growth and development. They have nearly reached a state of equilibrium. I say nearly, because for a time unfertilised ova, even of the higher animals, may undergo similar changes to those which fertilised ova undergo; although these changes soon become languid and end in decomposition. Their energy is in fact almost entirely potential. Hence, when left to themselves they disintegrate, and are absorbed or cast out. But by their union a fresh process of cell multiplication is excited, which is the beginning of a new evolution.

Evidently, then, fertilisation does not depend upon any special peculiarities of the sperm cells and germ cells.

How does it happen that some organisms multiply by homogenesis, and others by heterogenesis? Why cannot reproduction be carried on in all cases, as it is in many, by agamogenesis? What is the real significance of gamogenesis?

Spencer has asked and essayed to answer these questions as follows, premising that in the present state of biological science a complete answer is impossible.

From such considerations as have been previously

stated with regard to the relation between nutrition and reproduction, it follows that there is a certain relation between the commencement of gamogenesis and declining growth. We have seen that agamogenesis continues as long as the forces which result in growth are greatly in excess of the antagonistic forces; and that gamogenesis arises when there is an approach to equilibrium between these two sets of forces. Agamogenesis implies by its amount a large excess of nutrition, whilst gamogenesis implies by its amount a small excess of nutrition; the one or the other mode of reproduction occurs according as the external conditions are or are not favourable to nutrition. Generally speaking, only when growth is declining, do sperm cells and germ cells begin to appear; and the fullest reproductive activity arises as growth ceases. The "general law," says Spencer, "to which homogenesis and heterogenesis conform, thus appears to be that the products of the fertilised germ go on accumulating by simple growth, so long as the forces whence growth results are greatly in excess of the antagonistic forces; but that when diminution of the one set of forces, or increase of the other, causes a considerable decline in this excess, and an approach towards equilibrium, fertilised germs are again produced."

It cannot be said that a decrease of this excess always results in gamogenesis, for some organisms—

as the weeping willows—multiply only by agamogenesis, although the same causes of local innutrition are present in them as in other trees. All that can be said is—"That an approach towards equilibrium between the forces which cause growth and the forces which oppose growth, is the chief condition to the recurrence of gamogenesis; but that there are other unknown conditions, in the absence of which this approach to equilibrium is not followed by gamogenesis."

In seeking an answer to the second and third questions, Spencer points out that as gamogenesis recurs only in individuals approaching a state of organic equilibrium, and as the sperm cells and germ cells thrown off by such individuals are cells in which developmental changes have ended in quiescence, but in which after their union there arises active cell formation, it may be suspected that the approach towards a state of general equilibrium in such gamogenetic individuals is accompanied by an approach towards molecular equilibrium in these cells also. Hence, "the need for this union of sperm cell and germ cell is the need for overthrowing this equilibrium and re-establishing active molecular change in the detached germ"—a result which is probably effected by the mixing, in impregnation, of the slightly different protoplasmic substances of slightly different individuals.

I have already pointed out that the germ of every organism in its earliest stage is a simple protoplasmic cell, that it is by cell multiplication that all the early developmental changes are effected, and that the various tissues and organs of the developed organism are at first cellular.

It now remains for me to call attention to the further important fact, that cell multiplication is the means by which, throughout life, the wonderful processes of repair and reproduction of lost parts are carried out. It must be borne in mind that any group of cells, completely uncontrolled and placed in fit conditions, will tend to arrange itself into the form of the organism whence it originated. On the other hand "cells which form a small group involved in a larger group, are subject to all the forces of the larger group, will become subordinate in their structural arrangements to the larger group, will be co-ordinated into a part of the major whole, instead of co-ordinating themselves into a minor whole." Hence a small detached bit of a hydra soon moulds itself into the shape of an entire hydra; whilst the cellular mass which buds out in place of a lobster's lost claw gradually assumes the form of a claw, that is to say, it has its parts so moulded as to complete the structure of the organism. Thus the undifferentiated cells of the reparative new formation develop, not into new organisms, like the parental organism, but by a modification of their evolution they constantly reproduce the same tissues in the same parts, so that the form of the individual is maintained. We see, then, that there is no fundamental distinction between the processes of reproduction and repair; the latter is merely a modification of the former. Similarly, between the power which makes good an injury, or reproduces a lost part, and that which previously "was occupied in its maintenance by the continual mutation of its particles," there is no real difference. In the words of Paget "The powers of development from the embryo are identical with those exercised for the restoration from injuries; in other words, the powers are the same by which perfection is first achieved, and by which, when lost, it is recovered. . . . . It is one and the same power which being maintained continuously from the germ to the latest period of normal life, determines all organic formation."

Many organisms at certain periods of the year lose by a spontaneous and natural process certain parts of their body, which are subsequently renewed, *e.g.*, the fall of the stag's horns, the moulting of birds, the renewal of the cuticle of serpents, and of the shell of crustacea, and the fall of the leaves of trees. Similarly wounds heal, fractures are repaired, and lost parts are reproduced. Man and the higher animals possess this property in but a very limited degree, but its perfection in some of the lower animals is truly astonishing. The capability of repairing injuries and reproducing lost parts is not the exclusive property of living things, for even crystals manifest it. In fact, it appears that all bodies having definite form and construction possess the inherent power, under favourable circumstances, of repairing the injuries resulting from the action of external forces.

In order to account for this, we must suppose that by a kind of selective assimilation, the molecules of each part have the power of moulding the adjacent nutritive materials into molecules after their own kind. "Like units," says Spencer, "tend to segregate, and the pre-existence of a mass of certain units produces, probably by polar attraction, a tendency for diffused units of the same kind to aggregate with this mass rather than elsewhere."

In the case of the reproduction of a lost part, we must assume that the organism, as a whole, exercises some such power over the newly forming part, so as to make it a repetition of its predecessor. "If a leg is reproduced where there was a leg, and a tail where there was a tail, we have no alternative but to conclude that the aggregate forces of the body control the formative processes going on in each part." These processes may, therefore, be regarded as due to forces analogous to those by which a crystal reproduces its lost apex, when placed in a solution like that from which it was formed. In either case, as Spencer remarks, "a mass of units of a given kind shows a power of integrating with itself diffused units of the same kind; the only difference being that the organic mass of units arranges the diffused units into special compound forms, before integrating them with itself."

The power of repair and reproduction of lost parts varies in different beings: thus, although it is possessed even to the restoration of a lost limb among the lower animals, no such power is possessed by the highest. The general law to which these processes conform, is identical with that which we have seen holds for the process of reproduction. That is to say, it is greatest where organisation is lowest, and it almost disappears where organisation is highest. Hence we see it in perfection in those organisms which propagate the species by gemmation and allied methods. The gradual diminution of this power in the higher organisms is likewise due to the same cause which determines abatement of their reproductive power—the using up of the primitive germ protoplasm, in the formation of highly specialised structures, consequent on the physiological division of labour, and, in a less degree, in the growth and maintenance of structures already formed. This power is proportionately much greater in the young than in the old of all species, and it gradually lessens as life advances.

Seeing that the difference between the completeness of repair in early youth and adult age is so much greater than the difference in adults of different ages, we infer that the capacity is diminished more by development than by mere growth or maintenance. In other words, "to improve a part requires more and more perfect, formative power, than to increase it does."

Hence the larvæ of amphibians, which present many parts that in other animals are developed only in the embryo state, have also a greater power of reproducing lost parts than the perfect animals; and in the larvæ of insects lost parts are often reproduced, which in the perfect insects cannot thus be replaced.

This rule is equally true even for the most highly organised beings; thus, in the human species, cases have been recorded of more or less complete reproduction of fingers after injury, in early periods of embryonic life, though nothing of the kind occurs in adults.

Among the polypes, such as the *Hydra*, almost any minute group of cells separated from the perfect body suffices for the reproduction of the entire organism.

The well known experiments of Trembley have shown it to be a matter of indifference whether the hydra is divided longitudinally or transversely, or whether portions are merely cut out of its side; in all cases the separated fragments grow into perfect polypes. Even if the creature is cut up into a number of small pieces, each of these becomes a perfect hydra, and this process can be repeated over and over again with a similar result. Thus Trembley cut a hydra into four pieces, each became a perfect hydra, and whilst they were growing he cut each of them into two or three. These fractions being on their way to become perfect animals, he again divided, and thus he went on until from one hydra he obtained fifty, each of which developed into a perfect individual. Polypes produced in this manner grow much larger and are far more prolific than those which have never been cut.

Dugès has shown that *Planarians* possess the like property in a high degree. He has seen from eight to ten new individuals formed from sections of a single animal; and it has been observed that these creatures sometimes reproduce themselves spontaneously in a similar way. By dividing the anterior part of some of these animals longitudinally, Dugès succeeded in producing double monsters with two perfect heads.

In such cases it is quite impossible to discern any distinction between the processes of repair and reproduction by gemmation or fission. The identity of these processes is also shown by the fact observed by Trembley, that the supervention of gemmation retards the reproduction after injury.

In the higher *Actinozoa* half an individual will grow into a complete new being; and it has been noticed that when, after injury, the holothurians expel the whole of their viscera, these are in the course of a few months reproduced. Some of the lower annelids, such as *Nais*, may be cut into thirty or forty pieces, and each piece will eventually develop into a new individual.

As we ascend in the scale of organisation, this power, though much diminished, is still considerable. In insecta, arachinda and crustacea, entire organs, such as extremities, eyes, maxillæ, &c., may be reproduced when detached, but such separated parts never develop into new individuals.

Among the vertebrata in the case of *Fishes* fins are reproduced, but by far the greatest reparative powers are possessed by *Amphibians*. In many of these the reproduction of an entire limb or a tail occurs readily, and this even several times over, though with decreasing completeness. Spallanzani cut off the legs and tail of a Salamander six times, and Bonnett eight times, successively, and they were reproduced. Only the larvæ of the tailless Batrachians, as previously mentioned, and not the adults, are capable of reproducing lost members. In the Triton, Blumenbach has seen even the eyes, with cornea, iris, and lens, reproduced within the space of a year. But in the highest animals such complex parts are never reproduced; in these single tissues alone are regenerated.

In *Mammals* and *Birds* the only manifestation of this power is in the healing of wounds, which is often very imperfect in the case of the higher tissues. However, at a meeting of the British Association at Hull, in 1853, a thrush was shown, which was said to have thrice reproduced its tarsus after destruction by disease.

All highly specialised tissues possess but slight reparative power. Deficiencies of any size in muscles and nerves are filled up only with scar tissue. Severed nerves, however, are readily regenerated when the divided ends are kept in contact. Most epithelial and connective tissues have considerable reparative power; but cartilage is rarely replaced otherwise than by scar tissue.

Cases have been recorded of reproduction of supernumerary digits, after amputation, in human beings; this is remarkable, because, after amputation, the normal digits, it need hardly be said, have no such power of regrowth. The nearest approach to it is seen in the occasional reappearance in man of imperfect nails on finger stumps after amputation. But, according to Sir J. Simpson, in early embryonic life, human beings have considerable power of reproducing lost parts. He has several times observed that arms, amputated *in utero* by bands, &c., have grown again to a certain extent, and in one case the extremity was "divided into three minute nodules, on two of which small points of nails could be detected." These nodules clearly represented fingers in process of regrowth.

It seems probable that all the tissues of our bodies are constantly being destroyed and reproduced, in the normal exercise of their functions.

The same, or very nearly the same, power is seen in the case of plants. Thus a fragment of a begonia leaf, when placed under appropriate conditions, will develop into a young begonia, and so small is the fragment thus capable of reproducing a complete plant, that as many as a hundred may be produced from a single leaf. Various other plants manifest similar powers of multiplication, and there are well known cases in which a single epithelial cell suffices for reproduction of the whole.

Between these processes and the neoplastic tendency there is close affinity. They differ only in degree. It is well known that the tissues of repair are identical with those of neoplasms. The grand difference between them lies in this—that whereas the new growth of repair merely suffices to replace what has been lost, that of pathological neoplasms is indefinite, and knows no such bounds. In the latter case the normal subordination of the local processes to the specific hereditary tendency of the whole has ceased to exist. Hypertrophy is another closely allied kind of new formation. Here the newly formed tissue generally takes on the same structure and function as that of the part affected—physiological hypertrophy; but in many cases the hypertrophic new formation has no obvious function—pathological hypertrophy. These conditions are brought about by excessive growth and multiplication of the constituent cells of the affected part. Whole organs may be involved, or only certain parts. In the latter case there is gradual transition to tumour formation, as in the case of cutaneous warts.

Neoplasms are functionless redundant new formations, composed of histological elements similar to those of the structures whence they originate, but differing from these in respect to the time and place at which they appear. All of these processes are ultimately dependent upon cell growth and multiplication. As in the normal development, this is the initial step; and the succeeding changes are likewise modelled after the normal type. The grade of originisation attained by malignant neoplasms usually falls far short of that of the corresponding normal tissues whence they originate.

In the next two chapters I propose to treat of the evolution of vegetable and animal neoplasms, in accordance with these principles.

CHAPTER III.

THE EVOLUTION OF VEGETABLE NEOPLASMS.

"A pathological tumour in man forms in exactly the same way that a swelling on a tree does, whether on the bark, or on the surface of the trunk or a leaf, where any pathological irritation has occurred. The gall-nut which arises in consequence of the puncture of an insect, the tuberous swellings which mark the spots on a tree where a bough has been cut off, and the wall-like elevation which forms round the border of the wounded surface produced by cutting down a tree, and which ultimately covers in the surface—all of these depend upon a proliferation of cells just as abundant, and often just as rapid, as that which we perceive in a tumour of a proliferating part of the human body. The pathological irritation acts in both cases precisely in the same manner; the processes in plants conform entirely to the same type."
VIRCHOW.

WHEN a man begins to study any subject in a really independent spirit, the first thing he perceives is the connection between the various branches of science, by which they mutually enlighten and assist each other. This is especially the case with the subject we are about to consider—that of the evolution of neoplasms.

It appears to me that any attempt to give a scientific basis to our pathology of cancer and tumour formation must necessarily fail, which does not treat the subject in connection with *kindred processes* throughout the organic world, and consequently as part of a very large subject—that of Organic Morphology.

In advancing in this direction I have no predecessors to guide me with their critical light. If my task appears to be imperfectly performed, this must be my excuse. I will at any rate endeavour to give a right bent to inquiry.

We are indebted to the sagacious Johannes Müller, who continued the investigations commenced by Schwann, for insisting on the correspondence between the development from the embryo and the pathological neoplastic process. Virchow has further pointed out that the structural elements of all neoplasms are derived from pre-existing cells of the organism whence they originate. The diseased state is in fact merely a modification of the healthy state; or, in other words, the pathological process has its physiological prototype. But, strangely enough, no one has hitherto pointed out the physiological prototype of cancer and tumour formation. Consequently, current statements regarding the genesis and development of neoplasms are of the crudest and most conflicting nature. They reflect the prevailing confusion and its cause—the want of clear conception as to the nature of the morbid process.

This want I shall now endeavour to supply, by tracing out the evolution of certain vegetable neoplasms, in accordance with general biological principles. Although I shall in this chapter limit my remarks to vegetable neoplasms, the same treatment of the subject is equally applicable to animal neoplasms.

In order to make the matter clear, I will preface what I have to say with a brief account of the line of argument, which is but an epitome of what has been stated in the two preceding chapters.

Ever since the establishment of the Cell Theory it has been recognised that the life of every higher organism is but the expression of the collective functional activity of its constituent elementary parts or cells. Further, the one common factor underlying the evolution of all vegetable and animal structures is the development from cells; and each cell is a seat of life. Though each of the constituent cells of the higher organisms is to a large extent dependent upon others, yet at the same time each manifests a certain independence or *autonomy*. Every single cell may in fact be regarded as leading a kind of parasitical existence in relation to the rest of the organism.

We may go even further, since there are good reasons for believing that every component cell of the multicellular aggregates has the inherent power, under favourable conditions, of developing itself into the form of the parental organism: we may say then that each cell is potentially the whole organism. That the degree of development actually attained by each cell falls far short of this in most cases, is due to the restraining and modifying influence exerted by the whole organism on its protoplasm, which is thus compelled to the performance of comparatively subordinate, modified functions. In the performance of these special duties most of the original protoplasm is metamorphosed and used up. Hence in proportion as the cells are highly specialised their primitive reproductive function is either greatly reduced or altogether lost. But in the higher organisms certain cells never attain a high degree of development: they remain in a lowly organised condition, and serve, according as they are more or less *unspecialised*, either as germs for reproducing the entire individual, or for forming and maintaining the various tissues and organs. Such cells are found in all growing parts; they are the *only real cancer and tumour germs.*

I maintain that there is no fundamental distinction between the various modes of reproduction—sexual and asexual, the process of repair and formation of tissues, the reproduction of lost parts, and the different morphological variations, including bud, cancer, and tumour formations.

I regard all of these apparently so different phenomena merely as modifications of one common process which underlies and is the cause of them all, to wit, cell growth and proliferation. The particular outcome of the process in any given case can in the end be traced to the influence of the conditions of nutrition—understanding by this term the whole of the material changes wrought in the organism through its relations with the surrounding outer world. This being so, we can readily understand that under favourable conditions certain cells may take on independent action: may grow and develop without regard to the requirements of the adjoining tissues and of the organism as a whole. Thus the various morphological variations arise, whether anatomical or pathological.

In treating a subject from the evolutional standpoint, definitions are, as a rule, worse than useless. We have rather to acquire the habit of recognising the transmutation of related things into one another, so that our conceptions may harmonise with the sequence of events, as they occur in nature. To this end it is constantly necessary to refer to the past, and not merely to the immediate present. This is especially the case with the subject we are now entering on—that of bud formation in its protean aspects. In this connection—for the sake of brevity—none of the processes of gamogenetic reproduction will be included, notwithstanding the intimate relationship which these bear to the processes under consideration.

With this proviso I will now proceed to trace the subject genetically—both in individual cases (ontogenetically) and in the chief types (phylogenetically) —beginning with the study of the simplest forms, and proceeding successively to the most complex ones; for in this way we shall, I think, best arrive at an adequate perception of the essential relations. Indeed, it is impossible to demonstrate the nature and morphological phenomena in any other way. To understand the grown we must follow the growth; the history of development is the true light giver in every investigation into the nature of organic structure.

I think there is no doubt that the neoplastic process can be more satisfactorily studied in plants than in animals, owing to the absence, in the former, of many factors—such as the nerves and blood-vessels—which, in the latter, complicate and obscure the essential nature of the process.

In plants, the early embryo, and the lower animals, all the phenomena of growth go on without either nerves or blood-vessels. And so it is, at the outset, with all morphological variations, including bud, cancer, and tumour formations. In all of these cases the nerves and blood-vessels have not the slightest direct influence.

The lowest forms of vegetable life consist of single isolated cells, as in the *Protophyta*. In these lowly organisms these is no sexual reproduction: new individuals are produced by cell multiplication—by fission or gemmation—which is bud formation in its simplest form.

We are committed by the theory of evolution to the conclusion that at the earliest period of organic development these were the only kinds of buds; hence their formation must have preceded that of the higher and more complex kinds now extant. We may therefore infer, on phylogenetical grounds, that the primitive origin of every bud is a single cell. And ontogenetically, this is demonstrably so in the case of the lower plants, although in the higher ones it is generally impossible to trace the developmental process back to this early stage.

Such being the initial step in the evolution of bud formation, it follows from the law of heredity that the more complex kinds of buds, which have subsequently arisen, must have inherited their essential qualities from these simpler ones. Hence, we need not be surprised in studying these higher forms to find everywhere traces of this earliest phase of their evolution.

Advancing from the unicellular plants to the simplest multicellular ones, we see among the *Thallophyta*, the bud represented by a single cell or group of cells separated from some part of the body of the parent plant. Not unfrequently any part of the thallus may subserve this purpose. Such cells develop into new individuals *after* separation from the parental organism. Here growth exceeding a certain rate ends in the production of new individuals, rather than in enlargement of the old. We shall presently see the process undergoing various modifications in adaptation to other ends.

When, for instance, certain cells of the thallus or frond, instead of developing separately, multiply only in certain definite directions, whilst still maintaining their connection with the parent form, branched structures are produced instead of new individuals. This is the aspect of bud formation with which we are now more particularly concerned.

In seeking for the determining causes of the one or the other mode of development, it is to the conditions of nutrition that we must look. When the nutrition is relatively low the buds develop discontinuously; when it is relatively high their development is continuous; whilst changes in nutrition determine transitions from one mode of development to the other.

There are good reasons for believing that it is through the integration and development of proliferous outgrowths thus arising, that the higher kinds of plants have been evolved from the lower ones. Just as the thallus has arisen by the integration of cells, so we may conclude the higher plants have arisen by the integration of thalloid outgrowths into branching colonies.

Spencer has thus sketched the process:—"From fronds that occasionally produce other fronds from their surfaces, we pass to those that habitually produce them. From those that do so in an indefinite manner, to those that do so in a definite manner. And from those that do so singly, to those that do so doubly and triply through successive generations of fronds. Even within the limits of a sub-class, we find gradations between fronds irregularly proliferous and groups of such fronds united into a regular series. Nor does the process end here. The flowering plant is rarely uniaxial—it is nearly always multiaxial. From its primary shoot, there grow out secondary shoots of like kind." In the course of this evolution the distinction between axial and foliar organs has gradually arisen. We must, then, regard the simple proliferous outgrowths of fronds as the original of all kinds of bud formation in the higher plants. According to Spencer, the portion of one of the higher plants that corresponds to one of the primordial fronds is, "a foliar appendage together with the preceding internode, including the axillary bud when this is developed."

The details of the histological process by which thalloid bodies give rise to proliferous outgrowths or buds has of late been very accurately determined. On examination we find the growing point composed of a number of small, lowly organised, embryonic cells—

the *primary meristem.* Out of these cells the various kinds of tissue which subsequently arise are formed by differentiation. In most of the *Cryptogamia* the origin of the whole of these cells has been traced back to a single mother-cell, lying at the apex of the grow-

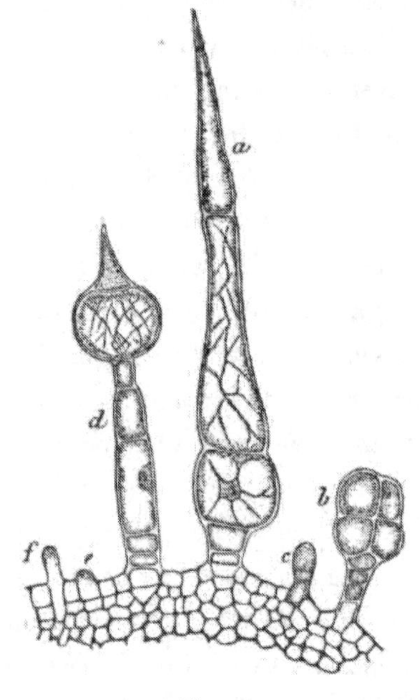

FIG. 7.
Illustrating the development of hairs.

ing point, the apical cell. This cell is usually larger than any of the adjacent ones. The primary meristem arises from its constantly repeated divisions in the following way :—two unequal daughter-cells are at first produced, of which the larger one remains from the first similar to the mother-cell, whose place it takes, developing into a new apical cell which repeats the process; whilst the other daughter-cell—the smaller one—appears like a piece cut off from the back or side of the apical cell, and is therefore called the segment.

In the simplest cases, when there is very little differentiation, the segment remains undivided; in such cases the structures produced by division of the apical cell take the form of simple rows of cells, as in some algæ, fungus hyphæ and hairs.

Let us pause a moment to follow out this interesting process as it occurs in hairs (trichomes).

Here we have (Fig. 7) a transverse section of the epidermis and subjacent tissue of the ovary of *Cucurbita* (× 100), showing hairs in various stages of development. From this it will be seen that each hair is the product of but a single cell of the external layer. In the earliest stages of development (as at *e*, *f*, *c*) each hair is nothing but a simple protuberance from such a cell. Some hairs never pass beyond this primitive uni-cellular stage. Others, however, grow into various forms in consequence of repeated cell-division, as above described. When, for instance, the divisions take place in one direction only, as transversely, a structure is formed consisting of a single row of cells; and when they take place longitudinally, the parts become thicker; whilst divisions proceeding at the same time in two or more directions give rise to masses of cellular structure which may develop variously (as at *a, b, d*).

In this way, then, hairs arise, and the process is a typical one. It constitutes the simplest and most instructive example of the formation of cellular growths—whether anatomical or pathological—to be found in the whole range of biology, and as such I particularly invite attention to it.

Outgrowths similarly arising from cells which lie beneath the epidermis originate the structures called emergences, such as the prickles of roses, which may be regarded as transitional forms between hairs and foliar organs. Abnormal developments in this direction may give rise to warty excrescences.

Generally, however, the process is more complicated than that just described. The segment, instead of remaining single, divides into two cells, each of which again breaks up into two, and by repetition of the process many times in the daughter-cells, a mass of protoplasmic cells is produced constituting the primary meristem. A simple case of this kind is shown in the subjoined figure (Fig. 8).

A branch of the thallome of *Stypocaulon scoparium* is here represented, with two branchlets, *x* and *y*, and the rudiment of a third branchlet *z;* all the lines indicate cell walls.

A very large apical cell (*s*) is shown divided by septa

(I^a, I^b), and thus giving rise to the segments which lie in a row one over the other. Each of these soon becomes divided by other septa (II^a, II^b). By the

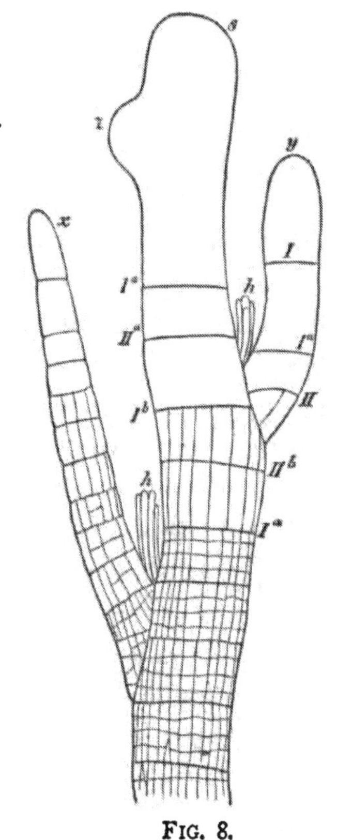

FIG. 8.
Illustrating growth with an apical cell.

formation, in each of these last of vertical, and afterwards of horizontal septa, numerous small cells arise.

Thus it is easily seen how the structure of the whole stem is built up of cells derived from a single segment, as at the lower part of the figure. In the same way each branch (*x*, *y*) originates as a lateral protuberance (*z*) from the apical cell; this becoming separated from the parent forms a new apical cell, which develops into its proper form by repetition of the above process.

Proceeding to somewhat more highly differentiated forms, we see in *Characeæ*—the lowest of the prothallus plants—buds giving rise to lateral outgrowths of the stem, which appear to be homologous with the leaves of the higher plants. The origin of each of these buds can be traced back to a single cell. The structure of the stem of *Chara* depends upon the properties of the cells which are successively derived by transverse division from the apical cell. Each segment thus cut off immediately divides again transversely into two superimposed cells, of which the lower one—the internodal cell—elongates greatly, but does not subdivide; whilst the upper one—the nodal cell—elongates little, but becomes greatly subdivided by successive vertical septa, forming a whorl of peripheral cells. It is from these cells that all the lateral members of the *Characeæ* originate, each from a single cell.

In still higher forms, such as the *Mosses*, *Equisetaceæ*, and many *Ferns*, leaves and branches originate in a similar manner, each from a single cell. Just in the same way the modified buds which develop into gemmæ, bulbils, detachable leaves and axes, and other forms of agamic reproduction common in these orders, take origin, each from a single cell.

From these types we pass through various transitional stages to the higher Cryptogamia and *Phanerogamia*, in which no apical cell can be detected at the growing point. There is no single primitive mother-cell recognisable as the original of the first rudiments of lateral structures, leaves, shoots, &c. These first appear as protuberances composed of a few or a considerable number of small embryonic cells without any definite arrangement (*k*, *k*, Fig. 9). At this stage, in their simple cellular structure, and in some other respects, these buds resemble the thalloid outgrowths of the lower Cryptogamia. But the cells soon group themselves into certain layers, which when traced backwards are found to be continuous with the epidermal tissue, the cortical structures, and the fibro-vascular bundles respectively; and may be recognised as the first rudiments of these structures. These relations are shown in Fig. 9, which represents a longitudinal section through the growing point of the stem of the embryo of *Phaseolus multiflorus*.

Covering the apex, we find an outer single layer of cells, the *dermatogen* (*s s*) continuous with the epidermis of the older parts; beneath this, a stratum consisting of several layers of cells, the *periblem* (*r*), from which the cortical tissues are derived, and enveloped by the periblem, a cylinder of somewhat smaller cells, the *plerome* (*v*), out of which proceed the axial tissues and fibro-vascular bundles. *pb* represents parts of two leaves, and *k*, *k*, their axillary buds.

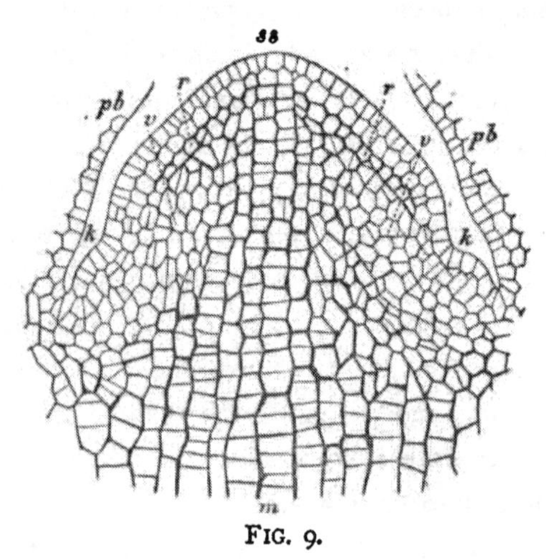

FIG. 9.

Showing the structure of the growing point without an apical cell.

It is evident, therefore, that whilst the various tissues and members of plants with apical growth have a common origin, it is otherwise with those which arise without apical growth, where each tissue-system is derived from its corresponding cellular stratum. It must not be inferred from this description that the processes in the growing point of

Phanerogams are essentially different from those in Cryptogams. The whole teaching of vegetable morphology is opposed to any such conclusion. The truth seems to be that our knowledge of the transitional processes connecting these extreme types of development is at present very defective.

Briefly, then, these are the conclusions at which we have arrived. The simplest buds consist either of single, lowly-organised cells, or of groups of such cells derived from single cells. The differentiated tissues of the parent plant have no share in the formation of the primitive bud, and have no connection with it until a later period.

Usually buds develop in organic connection with the parent stock; but sometimes they fall off and develop separately. Each bud is potentially a new individual, having under favourable conditions the power of developing into a perfect organism; but the degree of development actually attained often falls far short of this; buds become variously modified in adaptation to other ends.

In the higher plants, buds are usually distributed with great regularity, being either terminal or axillary; buds arising in any other way are said to be of adventitious origin.

Under ordinary circumstances a bud develops into a branch; but it does not therefore follow that all buds become branches. On the contrary, owing to disturbances in nutrition, buds may be very variously modified. Thus, in extreme cases, they may remain permanently undeveloped; or they may remain for long periods, even for years, in a dormant or latent state; and yet under favourable conditions their activity may revive.

When their evolution is somewhat less restricted the products may be dwarfed, or developed into spines, tendrils, or various irregular formations, which may properly be called tumours. Flowers, too, are merely modified branches.

Adventitious buds differ from normal buds only in respect to position; they originate in precisely the same way as those normally formed on the axis of plants. Adventitious buds have been found on almost every part of plants. They sometimes develop in extraordinary numbers on the stems and branches of trees, owing to some interference with the vegetation of the normal buds. Care must be taken to distinguish between true adventitious buds thus arising and old dormant "eyes," which, having been formed at an earlier period as normal exogenous buds, have been left behind and become enveloped by the subsequent growth. In a general way the root is distinguished from the stem by the absence of buds; but under exceptional circumstances adventitious buds may be formed even on roots. The physiologist, Duhamel, having planted a willow with its branches in the ground and its roots in the air, saw the roots become covered with buds, whilst the buried branches produced roots. The production of a flower-bud has been noticed on the root of a species of *Impatiens*. In some cases the divided root will reproduce the entire plant, as in the *Japan quince*, the *Osage orange*, and especially the *Paulownia*. In the last mentioned, even very small sections of the root will, when planted, develop into perfect trees.

There are many plants which produce buds on their leaves. This is of much commoner occurrence among the lower orders, and the ferns, than with the *Phanerogamia*. As examples among native plants may be instanced the *Watercress*, *Cardamine pratensis*, and *Malaxis paludosa*; in the last-named the buds are known to be derived from single cells of the surface of the leaves. *Begonias* exhibit this power in a remarkable degree; in some instances single scales from the leaves or stem habitually develop into young plants. One of the best known instances is that afforded by *Bryophyllum calycinum*, a succulent tropical plant, whose leaves produce buds furnished with roots, stem and leaves, at the extremities of its lateral nerves. These buds, which fall off spontaneously and root in the earth, may be compared to ovules which do not need to be fertilised before developing; and the leaf of *Bryophyllum* may be regarded as an open carpel, in which the seeds have been developed by nutritive action alone. Consideration of the subject in this light leads us to regard the bud as an individual vital centre, resembling the ovule. Many facts in vegetable physiology and pathology confirm this view. Schleiden regarded the ovule as a modified bud, and the now well-authenticated cases of parthenogenesis in plants support his theory. In these cases new individuals are developed from unimpregnated ovules, as from buds, the defect in the ovular reproductive power, which ordinarily renders impregnation necessary, being removed. The ovule or seed bud then differs from the branch bud essentially only in this: that it generally needs for its development the fertilising influence of the pollen. The fecundity of *Bryophyllum* completes the analogy between the bud and the fertilised ovule. According to Hofmeister, the buds of *Bryophyllum* arise before the complete unfolding of the leaf, as small masses of undifferentiated parenchyma in the deepest parts of the crenations of the leaves. Buds thus arising may take root and give origin to leaf-bearing branches whilst still in connection with the parent plant, as in *Drosera intermedia*, *Episcia bicolor*, &c. (Fig. 10), though they generally develop more readily after the leaves have fallen off. Adventitious buds have also been found on the petiole, lamina, stipule, and, in short, on every part of the leaf. It seems, indeed, as if *buds may arise wherever undifferentiated cells are present.*

Such anomalous bud formations in the higher plants may be ascribed to reversion of the cells to an embryonic state of activity; they remind us of the proliferous outgrowths so common among the *Thallophyta.*

The shoots which spring from buds usually develop into structures of the same form as those which compose the parent plant, but it occasionally happens that

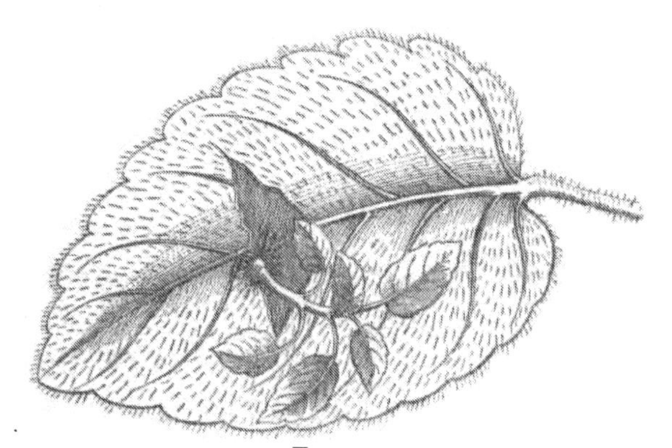

FIG. 10.
Represents an adventitious leaf-bearing shoot developed on a leaf of *Episcia bicolor.*

particular buds develop differently from others of the same stock. Gardeners call these changes "sports." Darwin has very carefully studied this remarkable process, to which he has given the name of bud variation.

Many cases of the kind are attributed by him to reversion; but he has determined that others can only be ascribed to the so-called spontaneous variability, such as often occurs in cultivated plants raised from seed. The common moss-rose is considered by Darwin to have originated in this way from the Provence rose (*P. centifolia*). Not only normal, but adventitious buds as well, are liable to this kind of variation. Fronds on the same fern, for instance, often manifest striking morphological deviations.

Of this a remarkable example is figured opposite (Fig. 11). It represents a portion of a frond of *Pteris quadriaurita*, in which the foliage emerging from an adventitious bud is seen to be very different from that of the rest of the plant. It accords with the theory of reversion that such anomalies are of much commoner occurrence among the lower than among the higher orders of plants.

In determining the nature of such variations, the constitution of the organism seems to play a much more important part than the direct influence of the environment. In illustration of this subject, Darwin says: "Let us recall the case given by Andrew Knight, of the forty-year-old tree of the yellow magnum-bonum plum, which has been propagated by grafts on various stocks for a very long period throughout Europe and North America, and on which a single bud suddenly produced the red magnum bonum. We should also bear in mind that distinct varieties, and even distinct species—as in cases of peaches, nectarines, and apricots, of certain roses and camellias—although separated by a vast number of generations from any progenitor in common, and although cultivated under diversified conditions, have yielded by bud variation closely analogous

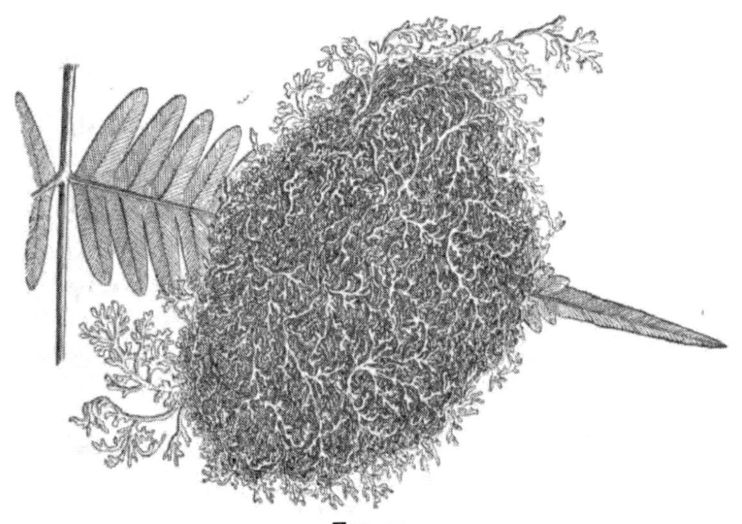

FIG. 11.

Portion of a frond of *Pteris quadriaurita* in which the foliage emerging from an adventitious bud is very different from that of the rest of the plant.

varieties. When we reflect on these facts we become deeply impressed with the conviction that in such cases the nature of the variation depends but little on the conditions to which the plant has been exposed, and not in any special manner on its individual character, but much more on the general nature or constitution inherited from some remote progenitor of the whole group of allied beings to which the plant belongs. We are thus driven to conclude that in most cases the conditions of life play but a subordinate part in causing any particular modification; like that which a spark plays when a mass of combustibles bursts into flame, the nature of the flame depending on the combustible matter, and not on the spark. No doubt each slight variation must have its efficient cause; but it is as hopeless to attempt to discover the cause of each as to say why a chill or a poison affects one man differently from another."

The formation of buds is then chiefly determined by the conditions of nutrition. Wherever there is an excess of nutritive materials, capable of being utilised for growth by the cells of the part, there buds arise. Under such circumstances buds may be formed wherever undifferentiated cells are present.

It has been observed in plants that any intermission in the local or general nutritive activity may determine the formation of buds. At first sight this seems to contradict a previous statement to the effect that buds may arise wherever nutrition is in excess. The explanation is, that this question of nutritive excess is a purely relative one; for we find that growth varies according to the surplus of nutrition over expenditure. In all cases bud formation implies a relative local excess of nutrition, for it occurs when the forces which result in growth are in excess of the antagonistic forces.

Thus we interpret the fact, that in so many cases bud formation takes place during the period of declining growth. And thus it is that plants may then be successfully transplanted, an operation which is generally impossible during the periods of active growth.

The buds which originate the various structures that appear at the beginning of every new period of vegetation, are formed at the expense of nutritive materials produced during the previous summer and autumn. At an early stage these rudiments pass into a state of rest before undergoing further development. But at the beginning of a new vegetative period they manifest renewed activity. It is the same with seeds, which, in addition to the undeveloped embryo, contain nutritive materials necessary for further evolution.

From the consideration of these interesting processes, we will now pass to a closely allied subject—that of the formation of Vegetable Tumours, which is essentially abnormal bud evolution. We shall find that in these cases the local changes are modelled after the processes of normal growth, and that both are subject to the same laws. Change of nutrition causing altered growth and impaired development seems to be the common ætiological factor underlying all of these abnormal formations. Their physiological prototypes must be sought in such kindred processes as those above described. The clear recognition of this important fact is likely to lead to great results in the future development of this branch of pathology.

My investigations into the nature of tumours in trees have led me to classify them into three main groups.

The first group comprises the discontinuous or circumscribed growths to which the vaguely used term of *knaurs* should be restricted. It includes all those

FIG. 12.
Showing five circumscribed tumours in the bark of a holly tree (natural size).

nodules so often met with in the bark of the beech, elm, oak, birch, holly, cedar, and other trees. Mr. Stephen Paget has lately presented some good examples of these tumours to the Hunterian Museum, where they may now be seen. They are generally single, but occasionally several are found close together. This was the case in one of Mr. Paget's specimens figured above—for which I am indebted to the courtesy of Mr. Eve—No. 546A in the general pathological series of the Hunterian Museum. It is described in the catalogue as a portion of the bark of a holly tree with the cambium layer and a small portion of the wood. Deep in the substance of the bark are five rounded tumours composed of hard wood, their cut surfaces showing faint concentric lamination. At the upper part of the specimen three of these have partially coalesced. They are distinctly circumscribed in the surrounding tissues; and there are no signs of any pedicles connecting them with the wood of the tree. The succeeding specimen, No. 546 B, shows a similar tumour from a beech tree, also devoid of any pedicle; and in No. 546 D we have five tumours from the surface, also of a beech tree, with one or two buds projecting from their surfaces. Here and there some of these buds have developed into shoots or minute branches.

It may be said of these tumours that they usually present as rounded or ovoid swellings in the deeper part of the bark, varying in size from a pin's head to a cocoa nut. The older nodules are generally found lying completely isolated in the bark, enclosed in a distinct capsule; a narrow fibro-vascular pedicle may sometimes be seen connecting the younger ones with the woody tissues of the trunk or stem. They are occasionally surmounted by a small stunted branch or shoot. On section, after removal, they are usually found to consist of very dense wood having a more or less concentric arrangement around a common centre. In most cases both pith and medullary rays can be seen. Sometimes, however, their woody structure is much more irregularly disposed. In short, they contain all the structural elements of the part whence they spring, but differently arranged. It would seem as if they were really shortened branches in which the woody layers had been abnormally developed in compensation for the curtailment in length and in other respects.

Lying completely isolated in the bark, these nodules increase in size at the expense of the surrounding nutritive materials. Hence, it often happens that their growth on the side of the cambium outruns that on the side of the free surface, where the nutritive supply is less abundant, so that a certain eccentricity in the arrangement of the laminæ is produced. Destruction of the bark which usually covers them in, whether as the result of injury or some other cause, such as pressure by the growing nodule itself, renders this tendency to increased growth on the side of the cambium still more obvious; for we then see complete cessation of growth at the exposed surface, which may even decay in consequence, whilst the side of the nodule next to the wood of the tree, owing to the proximity of the cambium, still continues to increase.

According to Dutrochet these tumours first appear as very small globular bodies in the cellular tissue of the bark, where they originate from unspecialised cells quite independently of the wood of the trunk, and this is the view generally held. Trécul, however, states that if they are examined at an early stage of development, they will be found to communicate with the woody tissue, whence they originate. The subsequent isolation he attributes to rupture of the primitive pedicles by the further growth of the tumours. Trécul also maintains that with rupture of their pedicles the nodules lose the power they at first possessed of originating shoots or branches; and when they retain this power it is due, he says, to the separation never having taken place.

Almost all observers are now agreed in ascribing the origin of these tumours to disorderly growth of adventitious or dormant buds. There is no doubt that buds may remain in a quiescent state for years, and then develop renewed activity under the influence of favourable conditions, the result being either a branch or one of these tumours.

Arising in this way we need not be surprised to find these tumours, like buds, possessed of a distinct individuality. It even occasionally happens that they may be used for purposes of propagation, as in the case with the "burrs" of some species of apples, which produce both leaves and roots in abundance.

In all of the above instances the result of the morbid process is the production of a highly organised circumscribed new growth, strictly comparable to the fibromas, lipomas, and other similar new growths of animal pathology.

Before entering on the consideration of the second group of tumours in trees, I must first of all offer a few observations with regard to certain intermediate formations of the nature of local overgrowths. I have often remarked in various trees the presence of hypertrophous branches; and in several instances I have seen localised hypertrophies involving the whole thickness of portions of the branches of such trees as the elm, lime, and oak. These conditions may be described as hyperostoses. They occupy in vegetable, as in animal pathology, an intermediate position between the processes of hypertrophy and tumour formation. They fall short of tumours chiefly in that they manifest but feeble signs of individuality. On this account I regard them as more akin to hypertrophies than to neoplasms. Such conditions appear to arise at an early stage of development, in consequence of excessive activity of the whole of the cambium at the affected part. Hyperostoses generally present as rounded protuberances covered with bark, their diameter being many times greater than that of the part whence they grow. On section they consist of hard wood, the concentric layers of which are unusually thick and very obvious, whilst the medullary rays, extending from the centre to the periphery, are but thin. Trécul has figured a very fine example of this kind from a birch tree.

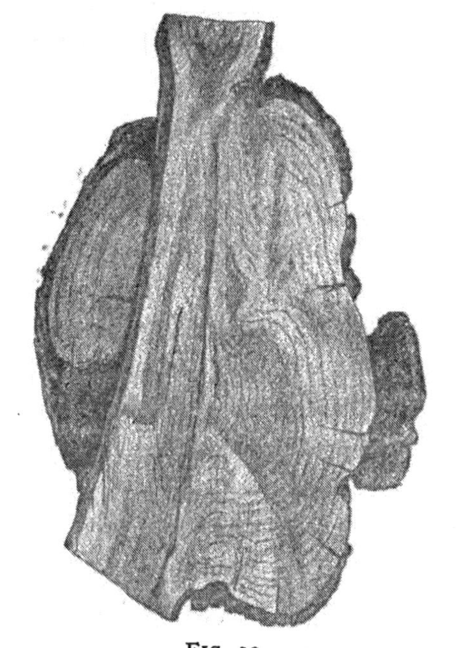

FIG. 13.
A continuous tumour (exostosis) from an elm tree, in longitudinal section.

The second group, comprising the continuous tumours—to which the term of exostoses should be restricted—present as woody outgrowths of the trunk or branch. They often attain great size, as in the example shown in Fig. 13, which weighed sixty-seven pounds, and measured sixteen inches by eleven inches. The diameter of the branch on which the tumour grew was four and a-half inches on the proximal and four inches on the distal side.

The general characters of these growths may be gathered from the subjoined brief description of this specimen, which was removed from one of the largest branches of a young elm tree—probably about thirty-five years old—on the main trunk of which was another similar growth, the tree being in other respects healthy. It was completely covered with rather hypertrophied bark, and two small branches grew from its surface near the periphery. On longitudinal section it was found to consist of very hard wood directly continuous with that of one side of the branch whence it grew. The laminæ composing the tumour showed a concentric arrangement of out-curved, irregularly wavy layers; whilst those of the branch itself appeared longitudinal (Fig. 13). Moreover, the former were many times thicker than the latter; in the broadest part of the tumour I was able to count twenty-six of these layers. A singular feature in this case is that a portion of the tumour had grown completely round the healthy part of the branch, outside the bark, and had then blended with the main tumour on the opposite side, so as completely to encircle the branch on which it grew.

Mr. Edwin Clarke kindly allowed me to remove this specimen from a tree in his meadow at Winchmore Hill. An adjacent elm tree of about the same size presented a single similar outgrowth on its main trunk. The situation was a damp one, and the soil heavy; close to the trees ran a small ditch, the water of which was highly contaminated with animal excreta.

Dutrochet has described these growths as arising in much the same way as the circumscribed ones —by a kind of bud formation or excessive local cell proliferation of the cambium layer; but their connection with the woody tissue of the stem is from the first more decided, and is never lost. These tumours must be regarded as abnormally developed branches. Such being their structure and mode of origin, they may be fairly compared with the exostoses and allied continuous tumours of animal pathology; and like them they are highly organised.

In the third group, I include those new growths which present a surface thickly studded with shoots and stunted branches. There is here a combination of the exostosis with diffuse bud formation. The French call such growths *broussins;* the term "burrs" might very well be employed as the

Fig. 14.
An old lime tree with an enormous growth densely studded with shoots and stunted branches.

English equivalent. Growths of this kind often attain enormous proportions.

An excellent example is figured on the opposite page (Fig. 14). The woodcut is from a photograph. It shows an old lime tree with a huge excrescence densely covered with brushwood and shoots, growing at the junction of the trunk with the main branches. Three other aged lime trees in the immediate vicinity were similarly affected. In all of these cases the general nutrition of the trees appeared to have suffered considerably, as indicated by numerous dead branches.

Fig. 15 gives an idea of the appearance of this kind of growth at close quarters. In the same neighbourhood I have seen similar growths on elm and birch trees. The trees of this particular locality are in fact remarkably prone to these growths, and to various other morphological anomalies, such as multiple excrescences, hyperostoses, and hypertrophies. Most of the trees are old and of large size. For a long period they have been left quite uncared for—innocent of the arts of forestry. The situation of the locality is peculiar. It forms part of a large shallow depression, lying at the slope of a hill. A large brook, the natural watershed of the neighbourhood, runs sluggishly through the low lying part. In its course it here forms several large lacustrine ponds, owing to the slight-

FIG. 15.
A near view of the growth represented in Fig. 14.

ness of the gradient. The water is highly charged with sewage matter. Clay formation underlies the whole neighbourhood, and forms the bed of the watercourse; whilst a layer of loose gravel rests on the clay. It results from this arrangement that the whole of the depressed area is thoroughly saturated with the sewage-contaminated water. The soil is heavy. A mist commonly hangs over the locality. As the result of these conditions its vegetation has long been abnormally nourished; even the grass shows it, for I suppose such large, coarse, rank grass could hardly be found elsewhere.

The peculiarity of growths of this kind, consists in the immense number of buds which develop adventitiously in the cambium layer of the affected part, where they either remain dormant or forthwith develop into shoots or stunted branches. The aggregation of these structures causes retardation and irregularity in the flow of the sap; whence the exuberance of the growth, and the curious markings produced in the wood. Moquin-Tandon mentions the case of a grafted ash in the botanic garden of Toulouse, where below the graft a large growth of this kind formed, from which proceeded over a thousand small, densely-packed, interlacing branches.

The production by these growths of large quantities of proliferating, lowly organised, cellular tissue, which subsequently undergoes imperfect evolution, constitutes the nearest approach in vegetable pathology to the sarcomas and carcinomas of animal pathology. The absence in vegetable tumours of anything like infectiveness (malignancy) may be urged as an objection to this comparison. To which the answer is, that in the absence of a highly specialised lymph-vascular system to transport the proliferous cells, such infectiveness is not to be expected; and that it is to deficiency in this respect, and not to any essential difference in the nature of the morbid process, that the absence of infectiveness in vegetable tumours is to be ascribed.

CHAPTER IV.

THE EVOLUTION OF ANIMAL NEOPLASMS.

"Our times may look with pride on the progress of modern morphology, whose importance is proved by the very fact that it is so destructive of previous views, and so fruitful in the most diverse directions." BILLROTH.

IN the preceding chapter I endeavoured to interpret the facts of pathological new formations in plants, in accordance with the doctrine of Evolution. I then claimed for the principles laid down, that they were equally applicable to Animal Neoplasms. It now remains for me to develop this part of the subject.

Underlying all I shall have to say is the great principle, that the evolution of pathological new formations is but a modified repetition of the normal evolution.

I propose then to treat of animal neoplasms in connection with *kindred processes* throughout the animal world. Hence, at the outset, the necessity of giving a brief sketch of animal organisation in general, as I believe it to be built up, both from the phylogenetical and ontogenetical standpoints.

Animals, like plants, consist either of single cells, of groups of cells, or of aggregated cellular groups variously modified.

Between the lowest forms of animal and vegetable life there is such close resemblance that no satisfactory line of demarcation can be drawn.

This has led some naturalists to unite all such lowest forms, whether animal or vegetable, into a separate kingdom—the *Protista*, intermediate between the other two.

I find it convenient to retain the name of *Protozoa* for the simplest forms of animal life, which consist of single isolated cells. These, like the *Protophyta*, have no sexual reproduction; new individuals arise by cell multiplication—by fission or gemmation, which is bud formation in its simplest form.

Hence, we may infer here, as in the vegetable kingdom, that the primitive origin of every bud is a single cell.

Among the *Foraminifera* a higher grade of organisation is foreshadowed by certain of them giving origin to buds, which instead of separating and developing into new individuals, continue to grow in connection with the parental organism, thus originating cellular communities. In these creatures, however, the integration is little more than physical;

each of the constituent cells still maintains its physiological independence.

A further advance in this direction insensibly leads to the multicellular animals—the *Metazoa*, which constitute the rest of the animal kingdom. Each of these animals commences its existence as a simple cell, similar to a protozoön, but instead of the adult state being attained by the direct metamorphosis of the protoplasm of the germ, as in the Protozoa, the first step in the development of all Metazoa is the conversion of the impregnated germ, by repeated cell division, into an aggregate of cells *(blastomeres)*; these then arrange themselves into layers *(blastoderm)*, whence the various organs of the adult bud forth in a regular and definite manner.

In such forms, although the component cells still retain a certain variable amount of physiological independence, yet they nevertheless remain united into one morphological whole, to which their various differentiations are mutually subservient. Thus regarded, each Metazöon is equivalent to a number of Protozoa, united to form a single individual of a higher grade of organisation.

It is a significant fact that in all Metazoa from hydra to man, the specialisation of cells which causes morphological differentiation begins with the formation from the blastoderm of a two cell-layered sac—the *gastrula*.

Among the lowest division of the Metazoa, the *Cœlenterata*, we find animals whose adult condition is analogous with this embryonic gastrula, which in the higher animals is only transitory.

As an example, I will instance the *Hydra*. This wonderful little creature is an aggregate of amœba-like cells, arranged in the form of a double-walled sac, with an opening at the aboral extremity leading into the digestive cavity. Recent researches have shown that its component cells are connected by strands of protoplasm often of extreme tenuity. The belief seems to be gaining ground that the component cells of most, if not of all organisms, are thus directly or indirectly connected with one another.

Though each hydra possesses distinct individuality, yet its component cells are so feebly integrated, that their lives are but imperfectly subordinated to the life of the whole. Each cell still retains to a large extent its physiological independence—its autonomy.

Between the cells of the outer wall *(ectoderm)* and those of the inner wall *(endoderm)*, there is obvious morphological difference. Yet when the creature is turned inside out the ectodermal cells so quickly take an endodermal function, and *vice versâ*, that the operation causes no apparent injury. Kleinenberg has described an interstitial stratum, composed of small indifferent cells, between these two layers; he regarded it as part of the ectoderm. Kölliker, who first noticed this stratum, thought it belonged to the endoderm. It seems probable, from various considerations, that part of it may be of ectodermal and part of endodermal origin.

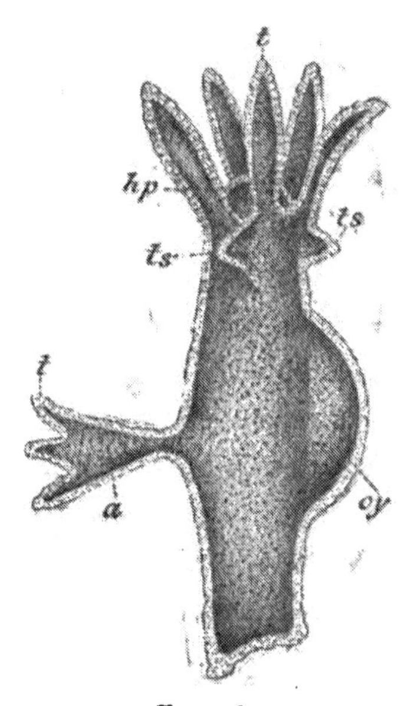

FIG. 16.

A large brown Hydra bearing at the same time an a-sexually produced bud (*a*) and sexual organs, ovary (*oy*) and testes (*ts*). *t*, Tentacles. *hp*, Hypostome.

In the development of hydra all the organs are produced by budding from these layers, and new individuals arise in precisely the same way (Fig. 16).

Thus tentacles, ovaria, buds and gemmiparously

produced individuals are all similar in origin. In fact the very same parts may be regarded either as organs or new individuals, according to their stage of development and the degree of dependence they hold to the parental organism.

In further illustration of the subject, it may be mentioned that the tentacles of hydra sometimes separate spontaneously and develop themselves into new individuals, and the like happens after they have been artificially severed.

Unfortunately we have no satisfactory account, as in the case of plants, of the details of the histological process by which buds originate in animals. The facts, however, with which we are acquainted relative to the development of animal tissues, place it beyond doubt that these buds arise like those of plants. In both cases the essence of the process is, that at certain points more intense cell growth and proliferation sets in than elsewhere. The groups of cells thus arising are at first similar, but they subsequently undergo special arrangement and differentiation. The precise nature of the resulting structures seems to be determined by the particular relations which each part has with the environment, either habitually in the individual, or occasionally in the race.

When a hydra is cut up into a number of small pieces, every one of them is capablo of developing into new being. The fragments having this reproductive power are so minute that it may be concluded, as previously mentioned, that every constituent cell is possessed of it. Therefore, it seems not unreasonable to ascribe the origin of the buds of hydra each to a single cell, although the process has never actually been traced to this early stage.

The minute anatomy of gemmation in hydra has never been thoroughly investigated. All that we know is that gemmæ first appear as minute roundish cellular prominences on the surface of the body, at any part of which they may form. These gradually elongate and assume a pyriform shape. At the free end a mouth is formed, and round it the tentacles sprout out. Then the young polype usually becomes detached and leads an independent existence. Thus a fresh hydra is formed by gemmation from its parent.

The development of the genital buds is better known. The two processes have so much in common that some light may be thrown on the former by studying the latter. Hydra presents the simplest possible condition of sexual apparatus, having no special organs for the male and female elements, which develop from cells of the body wall. Ovaria and testes are associated in the same individual. Throughout the greater part of the year hydra multiplies by gemmation; but in the summer pustule-like swellings appear on the surface of the upper part of the body, which were regarded by the earlier observers as the result of disease, until they were shown by Ehrenberg to be the true sexual organs. The testes are usually situated nearer the tentacles than the ovaria. It has not yet been definitely decided in which of the several layers the sexual cells originate. It has been observed, however, that at one or more points certain cells of the body wall take on more active growth than those adjacent; they enlarge and proliferate. By their repeated divisions collections of similar cells arise, which constitute the rudimentary ovaria and testes. In each ovarium a single ovum arises by one of the constituent cells, which usually occupies a central position, growing out of proportion to the rest, from which it is soon distinguished by its greater size. The essential elements of the testes arise from conversion of the germinal cells into small, clear, spheroidal bodies, each of which becomes a spermatazöon.

It not unfrequently happens in the gemmation of hydra, that a second generation of buds arises in the young polype, before its separation from the parental form. As many as nineteen young hydræ in different stages of development, have been seen thus connected with a single central polype.

In this way temporarily compound organisms arise. Such branching systems of individuals in many respects resemble plants; and, as with plants, under the influence of warmth and abundant nutritive supply, the process of gemmation may be made to continue almost indefinitely. The conditions of nutrition appear to determine whether the resulting growth shall be continuous or discontinuous.

With regard to the supervention of gemmation, we may in a measure account for it, by supposing that when the growth exceeds a given amount, there arises in certain small portions of the organism, an affinity of the molecules forming these parts for one another, greater than that of the different parts of the whole organism for each other.

In other divisions of the Cœlenterata, as the *Siphonophora*, we see groups of gemmiparously-produced individuals (persons), variously modified, permanently integrated into aggregates of a higher order (Fig. 17).

The component individuals of such a colony, which in other hydromedasæ lead separate and independent lives, are developed on a common contractile axis, around which the persons functioning as organs of the whole colony are arranged (Fig. 17). Here there are locomotive, nutritive, protective, tentacular, and generative persons, each having its special function and structure. Yet each person is but a modification of the common medusa type. This remarkable result is due to the division of labour very completely carried out.

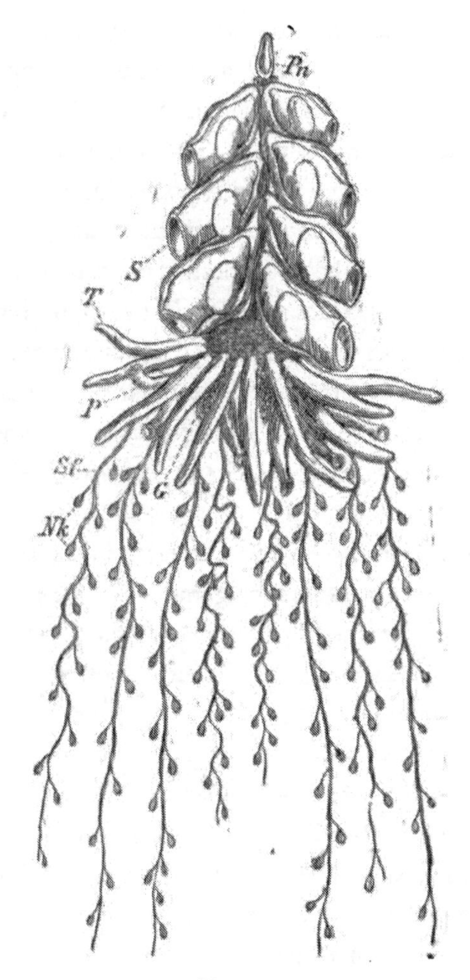

FIG. 17.

Physophora Hydrostatica. *S*, Locomotive persons (nectocalyces). *T*, Tentacular persons (dactylozoids). *P*, Nutritive persons (polypes) with tentacles *Sf.* *G*, Generative persons (gonophores).

The development of these creatures from the two cell-layered ciliated larva is instructive. It commences with a thickening of the epiblast at the future aboral pole, and the formation in this situation of a series of buds, which subsequently develop into the various persons. In the early stage, all of these various parts are alike, each consisting merely of a papilliform cellular outgrowth, but as development proceeds, each increases in size and gradually takes on its special form and function.

In the higher orders of animals the process of gemmation, in all its essential details, very closely resembles that just described. Therefore it will suffice briefly to run through its chief external features in these orders.

Among the many various forms of the important class of *Vermes*, the primordial outlines of the higher animal types may for the first time be discerned. Here we first of all meet with metameric segmentation and bilateral symmetry.

Each segment usually contains most of the organs essential to individual life; and when separated, either spontaneously or artificially, each is capable of independent existence. These creatures may be regarded as chains of gemmiparously-produced individuals, whose development has been impeded in consequence of the longitudinal integration. In the embryonic condition there are usually fewer segments than in the adult; new segments being budded off from the hinder end of the penultimate segment. Throughout the order reproduction by gemmation is common, especially among the lower orders. The process, however, is often confined to the larvæ, which differ from the sexually mature animals in form and habitation, and play the part of an asexual generation in the cycle of development, *e.g.*, *Tænia.*

It occasionally happens that the segments manifest the power of budding off *lateral* gemmæ, which supports the view that each segment was primordially an independent being.

With the *Polyzoa* gemmation from the wall of the body is a common process, and the buds usually remain adherent to the stock. In this way compound colonies arise by repeated gemmation from the primitively single embryo. In some species, however, the buds become detached. It often happens that all the persons of a polyzöon colony are not equally developed; hence we find here, as in the siphonophora, individuals functioning as organs.

The process of gemmation so common among the Vermes does not attain among the *Echinodermata*, except in so far as the animals themselves are the product of gemmation, and they have the power of reproducing lost arms.

The *Arthropoda* resemble articulated worms, in that they consist of a number of segments, the origin of which is attributable to disguised gemmation. In some of them the number of segments is increased by budding after leaving the egg, as with the higher annelida; but the process is generally much less complete, and the individuality of the segments is indistinct. In Vermes the organs are repeated in each segment; but in the Arthropoda they are common to the whole body. The most indifferent condition is in the Myriapoda, where the segments are similar and separate. In the course of the arthropod evolution a larval form (Nauplius) is developed, which is at first unsegmented, but subsequently segmentation arises through a gradual process of gemmation, which is nearly akin to the process by which the segmentation of the Vermes is brought about. Reproduction by budding is in this order limited solely to the generative system, where it is known as parthenogenesis. This phenomenon depends upon the germ cells forming by gemmation new individuals, independently of the influence of the male element. It occurs under two extreme forms; in one the parent is a perfect female, and the germs have all the characters of true ova; in the other the parent has incomplete genitalia, and the germs are deficient in some of the characters of true ova. In some instances the same individual is able to produce both ova and pseudova. This shows that the difference between gamogenesis and agamogenesis is not a very wide one. These processes may be regarded as links of a chain, which begins with parthenogenesis, and ends in alternation of generations. The multiplication by gemmation of the Cecidomyia larva is the result of an essentially similar process; but here the new formations arise at a very early stage of development, from the still indifferent germinal gland.

The adult *Mollusca* never present external segmentation, though indications of it may be made out in various organs. But at an early stage of development mollusca show distinct signs of segmentation, and of bilateral symmetry. This larval condition points to affinity with segmented organisms nearly allied to the Vermes. Among the mollusca, reproduction is never effected by any of those asexual methods so common in the foregoing orders. Here homogenesis is universal, and reproduction depends solely upon the activity of both sets of reproductive glands.

Of all invertebrate animals, the *Tunicata* are the nearest akin to the Vertebrata; they present also affinities with certain Vermes. In this class there is no well-marked segmentation of the adult body, though traces of it exist in certain regions. With the Tunicata gemmation is a very common process, and it often leads to the formation of colonies. It probably originated through division of the developing germ at a very early stage of its evolution. In the simplest case a bud grows from the body of the adult, and this bud eventually develops into a new individual, which becomes detached, and gives rise to other buds. Bud and parent reproduce sexually, as well as by gemmation—new colonies being formed from sexually produced embryos. In their next stage of complication the sexually produced larva gives rise by budding to a number of persons, which form the sexual generation. Here the sexually produced persons never themselves acquire sexual organs, but these organs originate in persons developed by gemmation. In this way a rudimentary form of alternation of generations arises. In Pyrosoma this is much more distinct.

Although in the *Vertebrata* there is no obvious external segmentation, yet in some systems, as in the axial skeleton, segmentation is exceedingly well marked. Spencer has argued in favour of vertebrate segmentation having mechanical rather than genetical significance. But the view now generally entertained is that the segmented vertebrate body has been evolved, as a secondary product, from an unsegmented primordial form, just as the segmented worms and the nearly allied arthropods have been evolved from an unsegmented worm form—by terminal budding.

Thus each segment, as in the Vermes, represents a greatly modified gemmiparously-produced individual, and the vertebrate body may be regarded as a longitudinally integrated series of such gemmæ.

Throughout the group homogenesis is universal; every vertebrate animal develops from a fertilised germ, and unites into its single individuality the whole product of such germ. Very exceptionally, however, cases arise in which there is departure from this rule, for at an early stage of its evolution the developing germ may divide, and originate two more or less complete individuals, *e.g.*, homologous twins and double monsters.

We may, then, recognise in the architecture of the higher animals two chief types.

In the first, the integration of the component segments is but slight. Creatures of this kind arise from a single primordial segment, which by gemmation repeats its own structure. The products of the process persist as well-marked persons conjoined into a linear or arborescent arrangement, as in the segmented worms and the siphonophora. Though such polymorphous stocks present the properties of individuals, they are morphologically but aggregates of individuals reduced to function as organs.

In the second type, of which the vertebrate body is an example, the component segments are never well developed; for through the action of a powerful synthetic tendency, the gemmation is retarded, so that the new segments make but a gradual and disguised appearance. Here the individuality of the primordial segment dominates from the first.

In the phylogeny of such organisms integration may appear at any time and synthetise the units, or by the operation of the converse tendency, the component units may acquire more or less of the independence of individuals.

Such is a brief account of the important part played by gemmation in the phylogeny of animals.

Spencer has very ably summed the matter up as follows:—"As among plants, so among animals. A like spontaneous fission of cells ends here in separation, there in partial aggregation, while elsewhere, by closer combination of the multiplying units, there arises a coherent and tolerably definite individual of the second order. By the budding of individuals of the second order, there are in some cases produced other separate individuals like them; in some cases temporary aggregates of such like individuals, and in other cases permanent aggregates of them, certain of which become so definitely integrated that the individualities of their component members are almost lost in a tertiary individuality. Along with this progressive integration there has gone on progressive differentiation."

If we regard the cells combining to form the higher animals as autonomous beings, possessed of morphological and physiological independence, we shall then see, although there is no such thing as true alternation of generations in the ontogeny of such animals, that nevertheless, as Haëckel has pointed out, a very complex alternation of the constituent cells does take place, which has a resemblance to it. The developmental cycle commences with the union of the male and female reproductive cells, whence the cytula or fertilised germ results, which by agamic multiplication originates the mass of similar cells called the morula. These differentiate into the various cells of the blastodermic layers. By further agamic multiplication the cells of these layers originate the very many generations of variously modified cells, whence the different tissues and organs arise. All of these polymorphic cell generations multiply agamically. Eventually, however, two of them differentiate sexually, forming the ova and sperm cells. By the union of these in the act of sexual reproduction the developmental cycle is completed. The reversion of the cells has led them back to their original starting point.

The only difference between this process and true alternation of generation, lies in the fact that in the former the products of agamogenesis remain in close contact with one another to form a multicellular organism; whereas, in the latter, the agamic products (persons) which represent the different generations, are separated and free. But the conditions prevailing in siphonophora show that this distinction is not of fundamental importance; for in these creatures, as we have seen, the same persons—widely differentiated by division of labour—remain united into one stock, that in other hydro-medusæ lead separate and independent lives.

When we examine the phenomena of gemmation and pathological neoplasia in this light, it is obvious that the essential thing in both cases is the interpolation in the developmental series of additional agamic cell generations, owing to excess of nutrition in these situations.

The fertilised germ reaches maturity through a vast number of changes. These are most marked during the early period of embryonic life, but they are also manifest, though in a less degree, long after birth. Between these embryonic and post-embryonic changes there is no essential difference.

The ontogeny of every higher organism presents a two-fold progress, proceeding *pari passu.*

On the one hand, there is continuous perfecting of bodily structure by increasing histological and morphological differentiation, whence the various tissues and organs result; and on the other hand, there is continual transition from lower and more general, to higher and more specific, types of organisation.

We are indebted to Baër for pointing out that these metamorphoses indicate roughly the changes of structure undergone by ancestors. Each individual organism, in the rapid and short course of its own evolution, reproduces the most important morphological changes through which its long line of ancestors have passed.

This important generalisation may be more fully stated as follows :—

"In its earliest stage every organism has the greatest number of characters in common with all other organisms at their earliest stages; at a stage somewhat later, its structure is like the structures displayed at corresponding phases by a less extensive multitude of organisms; at each subsequent stage, traits are acquired which successively distinguish the developing embryo from embryos that it previously resembled, thus step by step diminishing the class of embryos that it still resembles; and thus the class of similar forms is finally narrowed to the species of which it is a member."

In the Vertebrata, from man to fish, there is, as Baër insists, "one common type of formation, a certain sum of similar organs, which in the embryo state of all are met with in perfect similarity; but which during their development assume different forms in different classes of animals, and are even in some reduced to a state lower than that of the original type."

The development of the individual (ontogeny), is in fact an epitome of that of the race (phylogeny), conditioned by the laws of heredity and adaptation. This law is as true for the various parts and organs as it is for the whole organism.

In the course of organic evolution all parts are not equally developed, for whilst some are reduced and disappear, others arise and become highly developed.

In the majority of animals the most important morphological changes occur during embryonic life, as is the case with man, mammals, birds and fishes. But in many amphibians, insects, &c., very remarkable changes of form occur long after completion of the fœtal development. Thus the herbivorous tadpole is gradually metamorphosed into the carnivorous frog, by loss of gills, suctorial mouth, tail, &c., with new formation of two pairs of limbs and eyes, the substitution of pulmonary for branchial respiration, and other important internal changes.

In man and mammals such changes are more restricted, but instances occur in the development of the teeth, thymus, breast, uterus, skeleton, in the changes as puberty, &c. These and many other facts indicate the possibility of tissues remaining unchanged for long periods, and then taking on new phases of growth and development by reversion to the state of embryonic activity. With such changes the various pathological new formations must be associated; these depend essentially upon a kind of abnormal gemmation.

In the ordinary course of organic evolution, the growth and development of the cells proceed in a regular and orderly manner, in accordance with the specific hereditary tendency of the whole. But the process once started does not cease on account of irregularity, or because it is taking a wrong direction. Hence cells may arise at a place where they have no business, or at a time when they ought not to be produced, or to an extent which is at variance with the normal formation of the organism. In the embryo monstrosities and malformations are thus originated, and at a later period of development the various pathological new formations. These extremes graduate so insensibly into one another that it is impossible to separate them.

I will now proceed to give a short account of this interesting series of events, beginning with the subject of monstrosities and malformations, which has of late received but scant attention.

It is generally admitted that the mass of undifferentiated protoplasmic cells, which in the ontogeny of the higher animals results from the proliferation of the fertilised germ, occasionally manifests reproductive properties similar to those of Hydra—multiplying by a kind of gemmation—so that from a single developing germ two or more gemmiparously produced individuals may proceed. Here, as in the lowest organisms, growth exceeding a certain amount tends to the formation of new individuals, rather than to enlargement of the old one. The subsequent development of such gemmiparously produced individuals may be either continuous or discontinuous, as is ultimately determined by the conditions of nutrition. In short, the laws which govern the growth and reproduction of the lowest organisms are equally applicable to the early developing germs of the highest animals.

Thus when the division of the undifferentiated embryo into two equal parts is *complete*, and each of these develops into a new being, homologous twins are the result; and this is the only instancc of *reproduction* by gemmation in the highest animals.

Similarly, when the division is more or less *incomplete*, we get the various degrees of double monsters. The intimate connection between the processes of new formation and hypertrophy is shown by the fact, that in the absence of any fission whatever, under conditions which tend to the formation of *monstra per excessum*, the embryo may subsequently become enormously developed, so that giants arise.

However different in their several degrees of malformation, double monsters may be arranged in one continuous series. The most complete are those in which two bodies, nearly equally developed, are attached to one another only at a certain point. The locality and degree of the fusions present many variations. The usual points of attachment are the sacrum, sternum, umbilicus and head. Such monsters differ but little from homologous twins, and in both cases the similar individuals are invariably of the same sex.

It sometimes happens that one of the bodies of such double monsters exceeds its fellow in size and degree of development. We thus pass gradually to another class of these malformations, in which only one of the two fœtuses attains its full development, whilst the other is more or less stunted. The former are called autosites, the latter parasites, because they depend for nutrition upon the body to which they are attached.

Of these two chief types may be recognised, in one the parasite adheres externally to the autosite as a kind of appendage (*per implantationem*); in the other it is more or less included and overgrown by the tissues of the autosite (*per inclusionem*).

In either type the parasite may be variously developed; thus it may be well formed but diminutive, or it may be imperfectly formed, consisting only of part of the body, a single extremity, or even of nothing more than a confused mass of embryonic and other tissues. In short, whenever the development of the parasitic fœtus falls below a certain grade, it is represented only by tumour-like formations—the so-called *teratomata*—which consist of a great variety of different tissues, chaotically arranged, and in various stages of development. Such malformations are most frequently met with in the region of the sacrum and sella turcica. They represent abortive parasitic fœtuses; the attempt to form a double monster has failed through one of the twins developing at the expense of the other. Such teratoid products are more apt to take on the pathological neoplastic process than normal tissues, because they contain more embryonic tissue; and because a structure that has once departed from the normal is more likely to do so again than one which has developed normally.

Between these various kinds of malformations, as Vrolik remarks, "there can be traced such degrees and modes of deviation from singleness as pass without one abrupt step from the addition of a single ill-developed limb to the nearly complete formation of two perfect beings."

Developmental disturbances affecting the embryo somewhat later produce less extensive re-actions; and the later the disturbance sets in, the less extensive are its consequences.

Thus we get a class of monsters arising from fission—not of the whole embryo, but only of its axial structures. Such malformations are commonest at its cephalic end. They vary from mere duplication of the pituitary body, to the formation of two distinct and complete faces.

By the end of the third month of intra-uterine life the general form of the body and its members is well defined. After this period, developmental disturbances affect chiefly the rudiments of particular parts, and the resulting malformations are of a local nature, *e.g.*, supernumerary digits, mammæ, &c. When in the presence of developmental disturbances tending to such malformations, no fission occurs, the affected parts undergo excessive growth—thus the local congenital hypertrophies arise. Sometimes only single tissues are malformed, whence congenital angiomas, warts, moles, lipomas, &c.

It occasionally happens in the evolution of almost all parts and organs, that small portions of their germinal tissue become detached and remain isolated in the adjacent tissues. Such sequestrated fragments may remain quiescent, or they may undergo changes similar to those which characterise the parental tissues, *e.g.*, many dermoid cysts. It cannot be denied that pathological neoplasms are more likely to arise in connection with the cells of such isolated fragments, than in connection with those of the normal tissues, for the reasons previously mentioned when treating of the teratomata. But admitting all this, I fail to see any evidence that more than a small minority of neoplasms originate in this way. It is quite impossible to account for the cases of neoplasms growing from neoplasms on Cohnheim's hypothesis. I can see no foundation for the antithesis between embryonic and post-embryonic processes, on which Cohnheim has so strongly insisted. It appears to me that the very same cells, which are normally engaged in building up and maintaining the tissues of the body, are also the germs whence neoplasms originate.

Excess of developmental power seems to be the common ætiological factor underlying all of these conditions.

In investigating the local changes in neoplasms, it is impossible not to be struck by the fact that at every turn the phenomena met with have their counterparts in the normal evolution. Pathological new formations, like all other organic structures, ultimately depend upon the processes of growth and reproduction going on in the cells of the part whence they originate. Pathological and physiological cells are alike in their morphological and vital properties. Cell and nuclear division conform entirely to the physiological type—even as to the details of karyokinesis. In both cases the tendency of the newly-formed cells to revert to the parental type is perfectly obvious, and it is especially so when the cells become converted into tissues. Hence it happens that throughout the whole range of pathological new formations, no structures of new and specific type are to be found, but we everywhere meet with structures which resemble the physiological tissues both genetically and histologically.

In the ordinary course of organic evolution, the

process of cell multiplication goes on until the proper amount of structure for the needs of the organism has been produced, then it ceases within certain limits. During this process most of the original undifferentiated protoplasmic cells are metamorphosed into special structures, only a few remaining lowly organised and capable of further growth and development. In the healthy organism these cells subsequently concern themselves solely in the maintenance and repair of the established structures. In this quiescent state they continue throughout the whole life of the individual, unless aroused by abnormal conditions. There is evidently a definite law regulating the development of the tissues and organs in relation to each other and to the organism as a whole.

Though each of the constituent cells of the higher organisms is to a large extent dependent upon others, yet, at the same time, each manifests a certain independence or autonomy. If we wish to understand the nature of the special changes underlying pathological new formations, we must never lose sight of this important factor, the *autonomy* of the cells.

There are good reasons for believing, as I have previously mentioned, that every component cell of the multicellular aggregates has the inherent power, under favourable conditions, of developing itself into the form of the parental organism; we may say then that each cell is potentially the whole organism. That the degree of development actually attained by each cell usually falls far short of this, is due to the restraining and modifying influence exerted by the whole organism upon its protoplasm, which is thus compelled to the performance of comparatively subordinate modified functions. In proportion as the cells become thus specialised, they suffer more or less complete loss of their primitive reproductive power. But certain cells never attain a high degree of development: they remain in a lowly organised embryonic condition.

As long as such cells are subject to the normal restraining influence of the organism, they develop in accordance with the specific hereditary tendency of the whole; but when this influence is weakened or withdrawn their potential reproductive power may become actual.

In the ordinary course of the life of the higher organisms the only cells thus set free are the reproductive cells. But any cells abnormally emancipated may then grow and multiply more or less *independently*, regardless of the requirements of the adjoining tissues and of the organism as a whole. Lowly organised cells which thus arise, not being required for the building up of the parental structure, become superfluous, and are no longer retained directly under its controlling influence. In their development such cells soon assume an *individuality* of their own; there is manifest in them a tendency to the formation of a new aggregate rather then to enlargement of the old one—a new, partially insubordinate, centre of development has arisen. A certain degree of independence or individuality is the essential characteristic of every neoplasm, which, in *ultimate analysis*, must be regarded as the product of a more or less abortive attempt of certain cells, to reproduce a new individual by agamogenesis. There is then some truth in the remarkable saying of Paracelsus: "that in such a disease a man is himself and another; he has two bodies at one time, enclosed the one in the other, and yet he is one man."

In this way the various pathological new formations originate, by reversion of the cells which are usually engaged in maintaining the normal structures, to an embryonic state of activity. In all of these cases, as in the new formations of embryonic life, the subordination of the local processes to the specific hereditary tendency of the whole is lost or diminished, so that unspecialised cells then manifest their potential reproductive qualities, by taking an independent growth and development. In short, there is departure from the definite order, limitations, regular stages, and fixed periods of the normal evolution.

Since the structures composing morbid growths,

form by a process similar to that by which the normal tissues and organs evolve, the next step in our investigation must be to work out their evolution in association with that of the corresponding tissues and organs.

Every higher animal originates from a single cell—the cytula, or fertilised germ—by continually repeated cell multiplication. A solid spheroid of similar protoplasmic cells—the *morula*—is at first produced. These cells are so little differentiated that each of them probably possesses all, or nearly all, the properties of the original germ. Within the morula fluid collects, and the cells spread out on its surface in the form of a single limiting layer—the *blastoderm*. The cells of the blastoderm next differentiate into two layers, one of which becomes more or less completely enclosed within the other.

A two-layered hollow sac, open at its aboral extremity—the *gastrula*—is thus formed. The outer layer is the *Epiblast*—the primitive integument, which functions as a protective and sensory organ. The inner layer is the *Hypoblast*, which is nutritive in function, and forms the primitive digestive organ. Every metazoön, in the course of its ontogeny, passes through the gastrula stage, which is regarded by Haëckel as the ancestral form of all the metazoa. Between these two layers a third—the *mesoblast*—

usually appears, which probably originates from the primary layers by differentiation. As the proliferation proceeds, the cells of each layer become, as it were, the germs whence the various organs of the adult bud forth in regular and definite order. All observers are now agreed that each blastodermic layer originates only a certain series of tissues. It will suffice for our purpose to state that in the vertebrata each layer originates the following structures:—

The *Epiblast* is the source of the epidermis and its derivatives—nails, hairs, and the various integumentary glands, including the mammary glands. Epiblastic involutions form the lining membrane of the mouth (*stomodæum*) and anus (*proctodæum*), with their glandular derivatives. The pituitary body is of epiblastic origin. The epiblast also forms the cerebrospinal axis, the sympathetic and peripheral nervous systems, and it plays an important part in the formation of the organs of special sense.

The *Hypoblast* originates the whole epithelial lining of the digestive tract with its derivatives, including that of the air passages, lungs, liver, pancreas, and the glands of the alimentary canal. The bladder also is hypoblastic. The primitive origin of the pleuroperitoneal epithelium and of the genital cells is probably hypoblastic, although these structures now appear to originate from the mesoblast.

The *Mesoblast* gives rise to the connective basis of

all parts of the body, to cartilage, bone and blood-vessels.

His distinguishes only two germinal layers—the *Archiblast* and the *Parablast*. The former he regards as originating from the cells of the ovum, the latter from the white yelk. As to the correctness of this theory I must leave embryologists to determine; but I shall not hesitate on this account to avail myself of the classification of His, because, independently of theory, it is perfectly natural and very convenient.

The archiblast originates all the products usually ascribed to the epiblast and hypoblast, including the nervous and muscular systems; in short, all the structures of the body except the connective tissue series and the bloodvessels, which develop from the parablast.

Most completed organs include tissues derived from more than a single germinal layer.

In the whole course of its subsequent development, a derivative of one germ layer never develops a structure originally derived from another. Hence, after the differentiation of the blastodermic layers no wholly indifferent cells are ever formed; that is to say, no cells are formed like those of the morula, from which any structure might finally develop. Cells subsequently arising can only develop certain tissues, viz., those which are normally derived from the germ layer whence they originate. Thus derivatives of the archiblast always remain within this type, and never originate parablastic structures. In the development of pathological new formations the same law is observed. As in the normal development the cell derivatives of the blastodermic layers are never transformed into each other; so, under pathological conditions, no such metamorphosis ever occurs. There are among neoplasms no transitions from one type of tissue to another. The law of the specific nature of the tissues is everywhere obeyed.

Since the origin and development of pathological neoplasms follows a course homologous with that of the tissues in which they originate, we may classify these growths, like the normal tissues in association with which they develop, according as they originate from cell derivatives of the one or the other of the germinal layers. That is to say, they are either of *Archiblastic* or *Parablastic* origin.

Further, since all organic structures are primarily derived from cells, by a process of differentiation, it follows that a really scientific classification should be based on the degree of development attained by the cells in their upward progress to form tissues and organs. In accordance with this I divide each of the foregoing classes of neoplasms into two subclasses, the *lowly* and the *highly organised.* This classification has the great advantage of indicating clearly the *genetic relationship* between the normal and the pathological processes.

All *Archiblastic* neoplasms are modelled after the type of epithelial tissue. They consist of newly formed epithelial cells of epiblastic or hypoblastic origin, and this is their essential characteristic. These cells are derived by proliferation from the normal cells of the past whence they originate; and they always manifest a tendency to reproduce the parental type of structure. Inasmuch as many archiblastic neoplasms in the course of their evolution mimic glands of the body, they have been called *organoid.* Some parablastic elements are usually associated with archiblastic neoplasms, but these elements form an unessential part of the tumour. Most archiblastic new formations never attain a high degree of development; they exhibit, as it were, only the initial stage of the evolutionary process which they mimic. The result is a confused and indefinite mass of structure, consisting of proliferating epithelial cells ingrowing into the adjacent tissues. Such are the *Epitheliomata,* which is the term I use for all malignant epithelial neoplasms. As there are several varieties of normal epithelial tissue, so there are several corresponding varieties of epitheliomata; the differences depending upon the nature of the constituent cells, which may be either squamous, cylindrical or glandular. What I call glandular epitheliomata are usually named carcinomata. Some of these neoplasms, however, occasionally reach a rather high grade of organisation.

In other instances the newly formed cells undergo well-marked developmental changes, the result being new formations closely resembling the corresponding parental structures, as in the case of the *Adenomata*, *Cystomata* (neoplastic) and *Papillomata*, which are all of them non-malignant.

The *Parablastic* neoplasms may be treated in precisely the same way. All the members of this class are of parablastic origin; and they are modelled after the type of the embryonic or adult connective tissues, in association with which they originate.

In many instances only a low grade of organisation is attained; the new formations consist of structures homologous with corresponding embryonic stages of development, or but little removed from them. Such are the *Sarcomata* and *Myxomata*, which resemble immature connective tissue, and are more or less malignant.

Often, however, the evolution of the morbid structure is much more complete; then we get *Fibromata*, *Lipomata*, *Chondromata*, *Osteomata*, &c., which differ but little from the corresponding normal, fibrous, fatty, cartilaginous and osseous tissues, and are all of them non-malignant.

Of course all such classificatory arrangements are merely subjective conceptions, and all imaginable grades of transitional forms exist.

My classification may be briefly stated as follows:—

- I. ARCHIBLASTIC neoplasms.
  - 1. Lowly organised :—
    - *Epithelioma.*
      - Squamous.
      - Cylindrical.
      - Glandular.
  - 2. Highly organised :—
    - *Adenoma.*
    - *Cystoma* (neoplastic).
    - *Papilloma.*
- II. PARABLASTIC neoplasms.
  - 1. Lowly organised :—
    - *Sarcoma.*
    - *Myxoma.*
  - 2. Highly organised :—
    - *Fibroma.*
    - *Lipoma.*
    - *Chondroma.*
    - *Osteoma.*

Many neoplasms are malignant, that is to say, they have the power of reproducing themselves locally after removal or in distant parts; whilst others have no such infective properties. Does this imply specific difference between the two kinds of neoplasms? Certainly not. Between malignant and non-malignant growths there are so many transitional forms, that it is impossible to sever the chain otherwise than arbitrarily. Upon what then does the difference depend? We must recollect that before any neoplastic action can arise in a part, the normal restraining influence exerted by the whole organism upon its cells, in accordance with the specific hereditary tendency of the whole, must be relatively weakened, modified or withdrawn.

As I have previously mentioned, the evolutional outcome of cells thus emancipated depends chiefly upon whether the emancipation is complete, or more or less incomplete, and upon the grade of organisation of the emancipated cells. The natural tendency of every completely emancipated lowly organised cell or cellular group, when placed in fit conditions, is to form a new individual by agamogenesis. But in the higher animals the cells are never wholly undifferentiated, and the emancipation is always more or less incomplete; so that in these, various structural modifications result, instead of new individuals, such as new parts, new tissues and neoplasms.

Among neoplasms the degree of emancipation or individuality is very variable, and so is the grade of organisation. Highly organised neoplasms are but redundant repetitions of the structures whence they originate; and they are thus highly organised, because their cells are only slightly emancipated from the control of the parental structure. But cells engaged in the evolution of highly organised structures, suffer loss or impairment of their proliferous (reproductive) power, owing to their protoplasm being used up and converted into special tissue. This is just what happens in the evolution of non-malignant neoplasms, all of which are highly organised; and this is the reason why such neoplasms lack infective or malignant properties.

On the other hand, all lowly organised neoplasms are more or less malignant; and the most malignant are the most lowly organised. If it is asked why such neoplasms are lowly organised, the answer is: because the emancipation of their constituent cells, from the control of the parental organism, is much more complete than in the case of non-malignant neoplasms.

What then is the anatomical element upon which malignancy depends? We know that lowly organised cells and tissues, when dis-severed from their normal connections, may still grow and develop independently. The familiar process of skin grafting is a case in point. Here we have convincing proof as to the reality of the autonomy of the cell. Considering what is known of the phenomena of embolism, and of the transportation of particles of colouring matter after tattooing, there need be no difficulty in believing that much smaller particles, such as lowly organised neoplastic cells, may be transported with facility along such channels. It may be inferred from the smallness of the capillary network, that these transported particles must be exceedingly minute—a single cell probably in most cases. The germs which give rise to metastases are of this nature.

They are proliferous cells of the primary neoplasm, which, becoming detached, have been carried off by the lymph and blood vessels. The secondary growths are derived by repeated proliferation from such cells and their progeny, so that they must be regarded as the offspring of the primary neoplasm. This explains the great resemblance there always is between the primary and secondary growths, the significance of which it is impossible to ignore. Moxon has sagely insisted upon this. He says, "The first cancer which appears has a likeness to the part in which it appears, and the secondary cancers arising from it have the likeness of that first cancer; and those who doubt that they came from that first cancer must show us why they have that likeness." It well may be that normal cells, transported into normal living tissues, would not find everywhere a suitable habitat in which to develop. Failing this they would be very likely to perish, as Cohnheim has pointed out, in consequence of the metabolic changes going on in the part. But in the case of malignant neoplasms, the cells manifest abnormal inherent activity, and there are good reasons for believing that the constitutional condition has ceased to be normal.

Malignancy then depends upon the indefinitely sustained activity of lowly organised cells, which grow and multiply independently, without ever reaching a high grade of organisation. To such causes I attribute the distinctions between malignant and non-malignant neoplasms, and not to any essential difference in the nature of the morbid process.

Viewing the matter in this light, I can see no probability of there being any truth in the theory that malignant neoplasms are the outcome of general blood disease, due to the presence of micro-organisms, like tubercle and syphilis. Cancer cannot be inoculated, and it is not contagious. Its infectiveness is due to the autonomous power of its constituent cells, which everywhere tend to reproduce the parental type; whereas the lesions of phthisis and syphilis are inflammatory products, caused by the intrusion of foreign germs *ab extra*. The agency of micro-organisms is no more necessary to account for the phenomena of cancer and tumour formation, than it it is to account for the development of a tooth or a hair.

I think De Morgan came nearer the truth than anyone else, when, in the memorable debate on Cancer, at the Pathological Society in 1874, he said: "I can see no analogy between new growth, whether as innocent as lipoma or as malignant as cancer, and the products of true general or blood disease. From the first a tumour is a living self-dependent formation, capable of continual growth by virtue of its own power of using the nutritive materials supplied to it. Nothing like this is seen in any of the blood diseases."

Creighton and others maintain that the constituent cells of the primary and secondary neoplasms are not the direct descendants of the primary neoplastic cells; but that these *infect* the adjacent cells, by a kind of spermatic influence, and thus excite in them morbid action similar to their own. The observations upon which this theory is based are not convincing to my mind; and as it is opposed to biological principles and without parallel elsewhere in organic morphology, it may, I think, be discarded.

We are now in a position to work out more in detail the evolution of the normal tissues, in association with their corresponding pathological new formations. The process which leads to the formation of the tissues is essentially one of differentiation. In this process larger or smaller groups of cells acquire distinct characters by taking on definite order and arrangement. The new cell complexes thus formed, from aggregates of similarly altered cells and their derivatives, are the tissues. Tissues make up organs, and are themselves composed of cells. Organs may be defined as parts of the body which perform particular functions. The conception of an organ is therefore a relative one. For our purpose the tissues may be divided into two main groups—the archiblastic or epithelial and the parablastic or connective. In the former, cells predominate; in the latter, intercellular substance.

Epithelial tissues consist almost entirely of cells, which are united only by a scanty intercellular substance. In this and in some other respects, epithelial tissues must be regarded as less differentiated than any other tissues of the body. Epithelium represents phylogenetically, and, therefore, ontogenetically, the oldest kind of tissue. Epithelial cells vary much in form; the chief types with which we need concern ourselves are the squamous, the cylindrical and the glandular. Such cells, in one or several layers, cover the external and internal surfaces of the body and line its closed spaces. The life of all epithelial cells is short and transitory; constant change is the order of the day. Epithelial cells manifest considerable formative activity. The germ cells of all animals, which develop into male and female sexual cells, are nothing but lowly organised epithelial cells. In the laminated epithelia, the formative activity is usually confined to undifferentiated cells of the deeper strata, which keep up the requisite supply of new cells by constant growth and proliferation. Various solid and fluid substances result from the secretory activity of epithelial cells, *e.g.*, nails, hairs, horns, and the different fluid secretions. In this last case epithelial becomes glandular tissue. In the simplest instances individual cells of an epithelial layer become secreting cells, and function as glands, by transformation of their protoplasm into substance different from that produced by other adjacent cells. As an example, the well known goblet cells may be cited (Fig. 18).

FIG 18.
Goblet cells, secreting epithelial cells, or unicellular glands.

A gland in its simplest form is then merely a modification of a single epithelial cell. In association with this it is interesting to remember that, according to Goodsir, each acinus of the more complex glands consists at first of but a single epithelial cell.

When several cells enter into the formation of a gland, they become arranged in the simplest cases round a central cavity, which receives the secretion. Thus the least complex multicellular glands, consist merely of depressions of epithelium into the subjacent tissues. From this fundamental form the more complicated glands are derived by further epithelial ingrowths. This is accompanied in most cases by further differentiations of the constituent cells; a distinction arises between the true secreting (acinous) cells, and those of the duct (tubular), which connect the acini with the general indifferent epithelial layer.

The secretions of glands are formed at the expense of the protoplasm of their individual epithelial cells, which is metamorphosed into the fluid of the secretion.

The earliest products of the histological differentiation of the epiblast and hypoblast, are layers of epithelial cells which function as the primitive organs. The organs of the adult are evolved from these layers by further differentiations. Certain of their cells become as it were the germs, whence these organs bud forth in a definite and orderly manner. The essence of the process is that at certain points more intense cell-proliferation sets in than elsewhere. Solid bud-like processes of proliferous cells thus arise, which as they increase, grow into the adjacent tissues. The further development of the bud or initial cellular mass may be either continuous or discontinuous; it may spread as ingrowth, outgrowth, or sometimes as both. The ultimate form of each organ is only gradually attained through subsequent successive modifications.

These various types of normal epithelial new formations are the prototypes of corresponding pathological new formations. In all cases the morbid productions result from a modified repetition of the normal evolution of the parental tissue. It is interesting to remark, that in the earliest stage of their evolution neoplasms, like tissues and organs, are devoid of bloodvessels and nerves, which only make their appearance subsequently. The first step in the evolution of pathological, as in that of normal new formations, is excessive epithelial proliferation. Hyperplasia and *not inflammation*, is the starting point of every neoplasm. We have seen good reasons for believing that the primordial starting point of every organ is a single cell; and every neoplasm probably has a similar origin. This need excite no surprise when we recollect the wonderful reproductive properties of cells.

In illustration of these principles I propose to trace the evolution of certain normal structures, in association with their corresponding pathological neoplastic variations. For this purpose, I have in the case of epithelial tissues selected the *Skin* and the *Breast*.

The *Skin* of the vertebrata is a complex structure, formed by the union of tissues which, both morphologically and genetically, are quite distinct. It consists of two chief parts: the epidermis or superficial part, which is made up of epithelial cells derived from the epiblast; and the dermis or deep part, which consists of connective tissue derived from the mesoblast. The line of junction between epiblast and mesoblast corresponds, therefore, in the adult, with that between epidermis and dermis.

The epiblast, from which the epidermis is derived, consists of a single layer of more or less columnar cells. The epidermis of the lowest vertebrate animal, the amphioxus, retains this condition throughout life. But in all other vertebrata, the epidemis consists of two more or less distinct strata.

The superficial or horny stratum is composed of several layers of flattened horny scales; but the cells of the deep, mucous or malpighian stratum, are rounded or cylindrical in shape, and but little differentiated. Between these strata are cells of transitional forms. The lowly organised protoplasmic cells of the deeper part of the malpighian stratum are the seat of all the active changes that take place in the epidermis. These cells persist throughout life, and by their power of growth and proliferation, new cells are constantly being formed. The whole epidermis is thus being constantly renewed. The activity of these cells is also the source of all the glands and other structures developed from the epidermis.

The form of the constituent cells of the epidermis varies according to their stage of development. The youngest cells, those next the dermal papillæ, are of cylindrical form, and arranged in a single row perpendicularly to the dermis. These are succeeded by several rows of rather larger, rounded, or polyhedral cells, with finely serrated margins, among which pigmented cells may be found. Next follows a narrow stratum of non-serrated, flattened, nucleated cells, containing granules, which present indications of commencing horny transformation. As we proceed upwards the horny change becomes still more marked. Above the septum lucidum the nuclei disappear, and the cells are converted into flattened horny scales. These, arranged in stratified layers, form the superficial part of the epidermis, of which the surface cells are constantly being shed and renewed from below. Such is the remarkable series of transformations through which every epidermal cell must go to attain its complete development.

Immediately beneath the epidermis are a number of small conical elevations of the dermis—the papillæ. The epidermis also fills in the spaces between the papillæ, which are practically embedded in it. In the development of the papillæ the malpighian stratum of the epidermis plays an important part. It causes these structures to arise, by processes of its proliferous cells growing into the subjacent fibrous stroma.

The chief morphological feature which marks the transformation of simple epithelial hyperplasia into epithelial cancer, consists essentially in an abnormal, sustained repetition of the above process. In cancer of the skin, proliferous cells of the interpapillary region of the malpighian stratum grow downwards into the dermis; at first they form knobby projections, and then ingrowing columns

(Fig. 19), which often ramify deeply in the subjacent tissues.

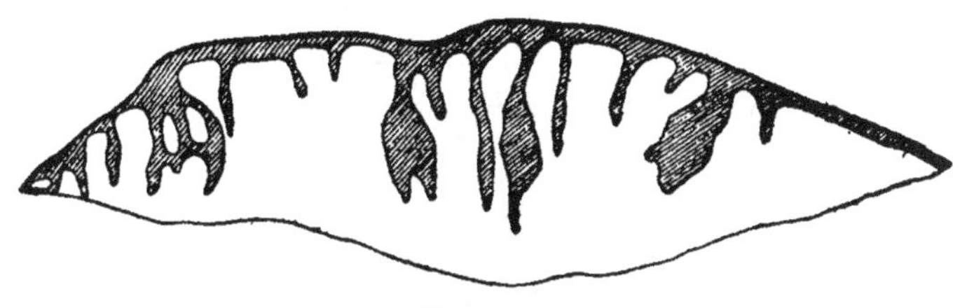

FIG. 19.
Section of the tongue on the confines of epithelioma, showing the irregular ingrowing of epithelial columns.

These cells grow and proliferate in semi-independence; that is to say, without regard to the requirements of the adjacent parts, and the organism as a whole. At first the cells of the ingrowing columns preserve their normal form and arrangement; but subsequently changes take place in their grouping, and "nests" are formed.

There is decided difference between cancers of the skin, according as the epithelial ingrowths enter the dermis more or less deeply. In some cases they seldom enter deeper than the cutis. The constituent cells of such ingrowths preserve for the most part their normal form and arrangements, and nests are of relatively rare occurrence. Such is the *superficial cutaneous cancer* or *rodent ulcer*. New formations of this nature progress slowly and are rarely infective.

In other cases—*deep cutaneous epithelial cancer*—the departure from the normal is more decided; the ingrowing columns grow rapidly and penetrate deeply, and sooner or later infect the adjacent glands and distant parts. The cells of such an epitheliomatous

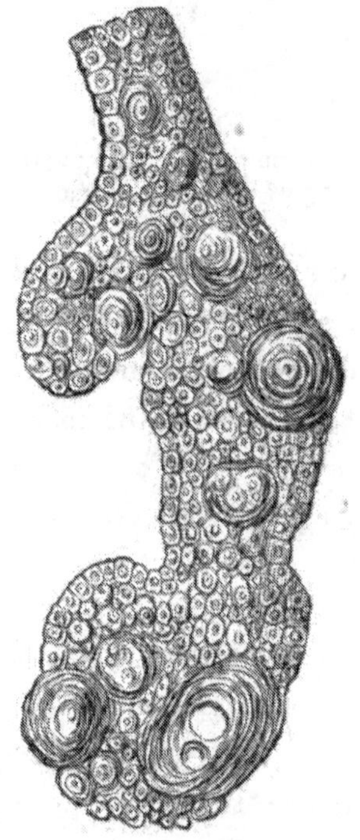

FIG. 20.

An epitheliomatous ingrowth, showing the "nests" an the outer layer of cylindrical epithelium.

ingrowth (Fig. 20), are epithelial cells like those of the epidermis whence they originate, and they retain the peculiar habits of the parental cells, including the remarkable one of undergoing a series of transformations in the course of their evolution. Although the arrangement of the cells of such an ingrowth bears a close general resemblance to that of the normal epidermis, yet it differs also in some important details. At the periphery we find a single row of cylindrical cells placed perpendicularly, as in the corresponding part of the normal epidermis (Fig. 20). Passing inwards we meet with several layers of rather large, rounded, nucleated cells, with finely serrated margins. According to my experience, these are much more numerous in pathological ingrowths than in the normal epithelium. In the centre we find flattened, non-serrated, hornifying cells, in the midst of which are the so-called "nests." These are rounded aggregations of laminated horny scales, having at their centre the cell or cells, from whose proliferation these structures are exclusively derived. It is easy to understand how such structures are formed by the irregularly evolving cells, when we remember the similar transformations that occur in the course of the regular evolution.

In the cutaneous *papillomata*, we have examples of much more highly developed epithelial new formations, with but slight departure from the normal type. All of these neoplasms are of an innocent nature. They arise in consequence of hyperplastic processes in the cells of the malpighian stratum. A thickening arises at the seat of disease. The interpapillar processes of the hyperplastic epidermis encroach somewhat on the subjacent dermis, which also becomes more or less thickened. In these cases the normal arrangement and grouping of the cells is maintained; but their power of growth and multiplication is beyond the normal requirements. This latter quality indicates feeble individuality, and points to affinity with other pathological neoplasms; but in

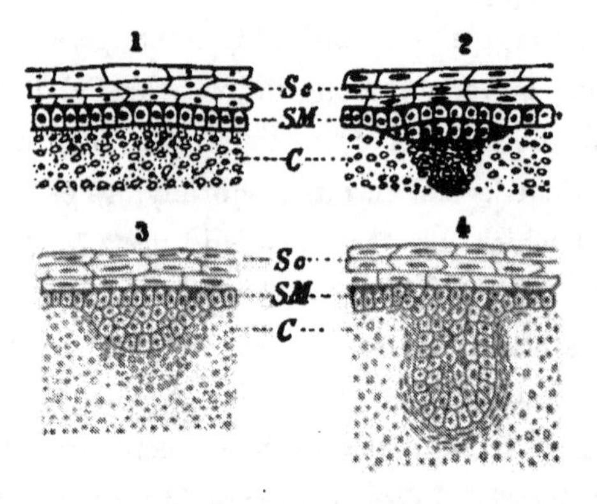

FIG. 21.

Diagram showing early stages in the development of the breast. *Sc*, Horny stratum of epidermis. *SM*, Malpighian stratum. *C*, Dermis.

other respects papillomata resemble local hypertrophies. Such then are the three chief types of cutaneous epidermic neoplasms. All imaginable grades of intermediate forms are met with. The gravity of any particular neoplasm will generally be found to depend chiefly upon its grade of organisation.

The human *Breast* is generally regarded as a greatly enlarged and modified cutaneous sebaceous gland. Like all other glands opening upon the free surface of the body, it is developed from the epidermis by a kind of budding (Fig. 21).

The process begins at about the fourth or fifth month of intra-uterine life, by certain cells of the malpighian stratum, in the site of the future organ, proliferating more rapidly than those adjacent (Fig. 21). A solid knob-shaped mass of proliferous epithelial cells, ingrowing into the subjacent dermis, is the result. A few weeks later, by repetition of the above process, secondary buds arise from the primary ingrowth, and likewise grow into the dermis. At this stage the rudimentary gland consists of a single branching system of proliferous epithelial cells, ingrowing into the surrounding tissues. These form the first rudiments of the ducts and lobes, and it is only subsequently that they become excavated.

Towards the end of intra-uterine life, each lobe develops a single external opening or duct, whilst at its cæcal extremity rudimentary lobules begin to appear as small cellular knobs.

At birth the organ consists of from twelve to fifteen lobes, the excretory ducts of which are excavated and lined with a single layer of small cubical cells, the rest of the organ being still solid. The knob-shaped lobules now present small buds, which are the rudimentary acini. These are aggregations of small irregularly shaped, nucleated epithelial cells: they constitute the matrix material for further development.

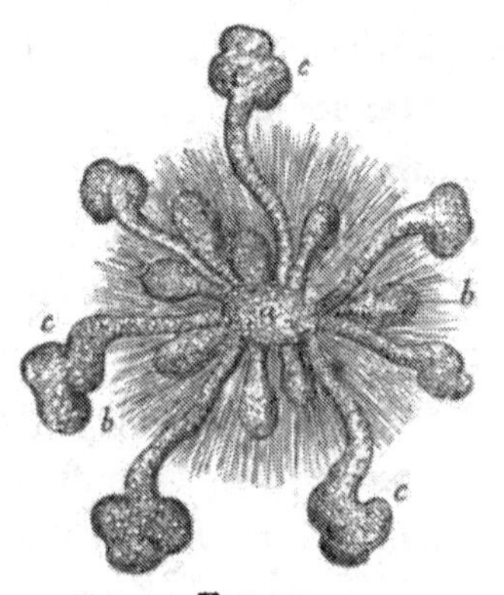

FIG. 22.
The mammary gland of a mature fœtus. Primary ingrowth (*a*) with secondary (*c*) and tertiary (*b*) offshoots.

From this simple condition, the permanent form is gradually evolved by further ramifications, which are but superinduced repetitions of the initial process. Thus the whole organ is gradually built up by the continuous extension of ingrowing, branching epithelial columns, which subsequently become excavated and otherwise perfected. In the course of its ontogeny the breast reproduces those simpler conditions, through which it must have passed during its pylogenetical career.

Soon after birth, in both sexes, the organ becomes swollen and tender, and a little milk sometimes escapes from the nipple, which is probably the product of the disintegration of the cells of the recently excavated ducts. But this soon subsides, and until puberty there is little further structural change—the true secreting tissue (the acini) remaining undeveloped in both sexes.

The male mammæ, being functionally inactive, usually continue in this imperfectly developed condition throughout life, although they experience slight temporary excitement at puberty. In man and some other male mammals, these organs occasionally become well-developed and secrete milk. Darwin accounts for the male mammæ being so much more perfectly formed, than the rudiments of other accessory reproductive structures found in one sex, by supposing that; "during a former prolonged period, male mammals aided the females in nursing their offspring, and that afterwards, from some cause (as from production of smaller number of young) the males ceased to give this aid; disuse of the organs during maturity would lead to their being inactive, and this state would be transmitted to the males at the corresponding age."

In the female breast remarkable structural changes

set in at puberty, although it is not until after pregnancy that the organ attains its full development. Before puberty the breast consists chiefly of excretory ducts; but as puberty approaches, the true secreting tissue arises, by the abundant new formation of glandular acini. The commencement of this mammary development usually precedes the first catamenial period, and at every subsequent period more or less temporary sympathetic reaction is excited in the breast.

But the most important mammary changes are those induced by the stimulus of conception, which converts the previously functionless organ into an active milk-secreting gland. During this period the acini attain their highest degree of structural perfection. This, however, is but a transitory condition, which ceases when the stimulus is removed, and is again excited by its repetition.

During the intervals between these periods of functional activity, the breast remains in a quiescent, functionless state—the resting stage.

The secretion of the breast is formed after the same type as that of a sebaceous gland, of which the following is a brief description:—Within the membrana propria of its secretory part, we find a stratum of small irregularly-shaped epithelial cells, each with a large nucleus (Fig. 23 *b*). The cells of this region are constantly proliferating, and as the products of the process gradually shift towards the duct, they become changed and gradually form the secretion. The steps of the process are as follows :—The cells next the marginal cells (Fig. 23, *b*), increase in size and their nuclei dwindle. As they approach the centre of the acinus their nuclei disappear, and the cells become distended with granules and oil globules.

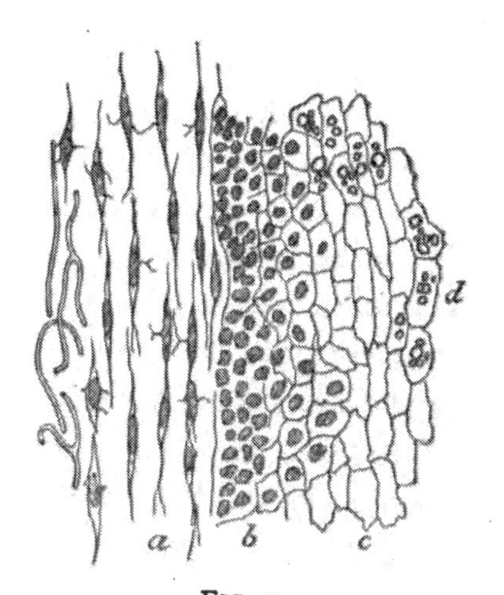

FIG. 23.

Section of the wall of a sebaceous cyst. (*a*) The fibrous stratum with embedded connective tissue corpuscles. (*b*) The marginal stratum. (*c*) Hornifying cells. (*d*) Sebaceous cells.

Finally they burst, and their debris forms the secretion, which is discharged. These changes may be compared with the analogous transformations of the cells of the epidermis, which eventuate in desquamation. Lactation is the outcome of a similar process.

Milk may be regarded as the product of the complete deliquescence of successive generations of epithelial cells, which are destroyed in the process, and replaced by relays of new cells, derived by division from other still active epithelial cells, of the part. Although the mammary glandular epithelium forms but a single layer, yet its cells, like those of the epidermis, are of a very transitory nature.

When in a sebaceous gland, the metamorphoses of the cells are from any cause unduly delayed, the constant proliferation going on in the marginal stratum causes great accumulation of imperfectly changed cells to arise, instead of the proper secretion (Fig. 23). It is under such conditions that sebaceous cysts originate. Here, the formative activity of the glandular cells, predominates over their secretory activity. We shall presently see that abnormal processes of this kind, going on in the mammary acini, are important factors in the origination of neoplasms.

The complete degree of mammary function, which eventuates in lactation, is only attained periodically, and the process is always gradual. The following is a brief account of Creighton's description of it. Subsidence of function goes hand in hand with undoing of structure, and revival of function with the building up of structure. Variations of intensity in the secretory force are measured by its products, which correspond to changing aspects of the secreting acini.

The beginning of the rising function coincides with the beginning of pregnancy, and the process occupies the entire period of gestation.

In the *resting stage* the gland is shrunken and surrounded by a considerable quantity of fibro-fatty tissue. The acini are shrivelled up. On microscopical examination of sections of the gland in this stage (Fig. 24), each acinus appears as an alveolar

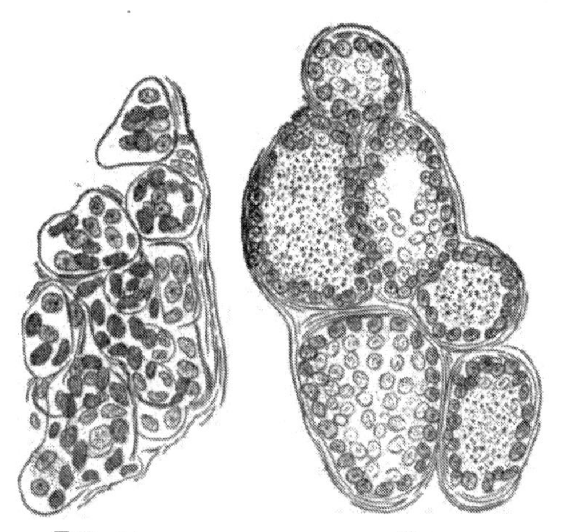

FIG. 24.
Mammary lobule near the resting stage.

FIG. 25.
Mammary lobule of rising function, not many stages from lactation.

space bounded by a thin layer of fibrous tissue, denuded of epithelium. Its contents are a few, irregularly arranged, epithelial cells, with large nuclei and scanty surrounding protoplasm.

During the *rising function* the size of the acini gradually increases from that of the resting stage. The cells increase in number and size, and acquire more protoplasm. They gradually arrange themselves so as to form a lining membrane for the wall of the acinus (Fig. 25); which, as lactation approaches, is converted into a regular mosaic. The cells become granular, irregularly shaped, excavated and vacuolated, secreting granular and mucous fluids. The milk of the first few days is always somewhat crude, containing colostrum cells, which are the last of the long series of secretory products thrown off during the period of the rising function.

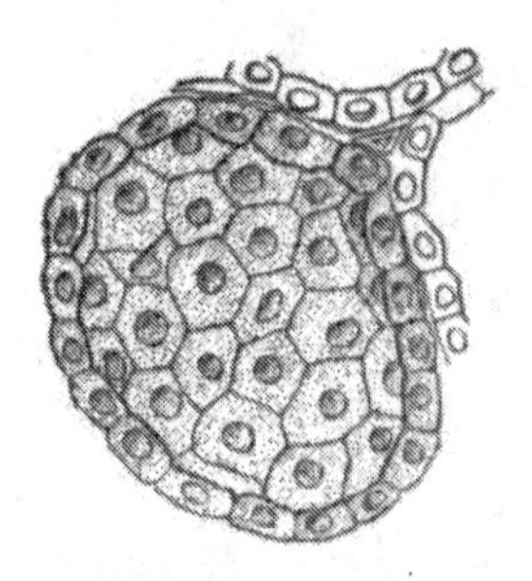

FIG. 26.
Fully expanded mammary acinus, showing the epithelial mosaic.

The *fully expanded acinus* (Fig. 26), in a state of active secretion, is at least four times as large as that of the resting stage. Its contained cells are much more numerous than at any other period; and they form a perfect mosaic, lining the membrana propria. Each cell is flattened and of polyhedric shape, and has a large nucleus surrounded by a broad zone of protoplasm.

During the period of *subsiding function*, the organ gradually returns to the resting stage through the converse series of changes. In this process the cells pass through a series of transformations, from the forms which characterise the perfect mosaic of lactation to those peculiar to the various stages of the subsiding process. These changes are accompanied by constant destruction and renewal of the participating cells.

It will be gathered from this description that the changes which take place in the breast, during the periods of functional activity, are but a sustained repetition of those of its embryonic development.

"The embryonic cells, in order to become secreting cells of the mammary acini, go through a cycle of changes; and the changes that they undergo are precisely those that the cells of the mature organ undergo in producing the periodical secretion."

It now remains for me to shew that the pathological neoplastic processes of the breast are due to a modified repetition of the same process. During the periods of rising and subsiding function the secretory metamorphosis of the cells is incomplete, so that instead of milk, only *cellular* waste products result. Creighton has done excellent service in pointing out how such cells become the germs, whence cancer and tumour formations originate.

He says: "Taking the breast at the resting state, it cannot under any circumstances reach the perfection of its function without going through the somewhat slow series of unfolding changes. When the evolution that is set up is of a spurious kind, or in other words, when the gland is disturbed from its resting state by some cause other than pregnancy, the steps of its unfolding are less orderly than in the normal evolution; and the fatality of the morbid process consists in this, that the spurious excitation never carries the gland to the end of its unfolding, or to the perfect function. The products of the gland never get beyond the crude condition, and it is the crude cellular kind of secretory product that makes the tumour."

With the decline of reproductive power, at the climacteric period, retrograde changes set in, and the breast suffers effacement of its secretory mechanism and withdrawal of its secretory force. The acini, and even the infundibula of the ducts disappear, so that eventually nothing is left but the ducts surrounded by dense fibro-fatty tissue. The extreme hardness of cancers of this period is probably largely due to the density of this fibrous stroma, in which the elements of the disease are embedded.

These climacteric changes may easily assume a

pathological character. Fragments of structure with correlated functional force, surviving the general obsolescence, are the sources to which we must look for the origin of the post-climacteric neoplasms. According to Creighton, the kind of cell most commonly found in such tumours of the breast is the large nuclear cell, with scanty surrounding protoplasm. Measured on the physiological scale, these cells belong to the intermediate stage of the unfolding process; they stand for a half roused physiological stimulus. The morbid force delays at this intermediate stage. The result is the accumulation of cellular waste products, instead of true secretion. The formative activity of the cells predominates over their secretory activity. Cells that should have passed out of the gland as waste products, remain at their place of origin; they proliferate and aggregate more or less independently. It is upon such deviations from the physiological track that the origin of mammary tumours depends. This accumulation within the acini causes them to become greatly enlarged (Fig. 27). Solid bud-like cellular processes arise from their walls, which grow and ramify in the adjacent tissues. All kinds of mammary epithelial neoplasms begin in this way. In the case of *Cancer*, the process seldom advances beyond this low grade of organisation. The crude formation is simply repeated and reproduced indefinitely. The morbid product, however, always has a certain likeness to the tissue whence it originates. Billroth describes two forms of mammary cancer, the acinous and the tubular. In the

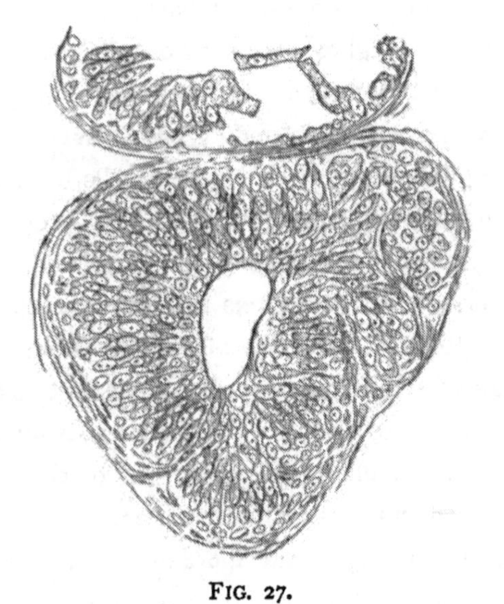

FIG. 27.

A pathological mammary acinus. The epithelium massed in several layers.

former large lobulated nodules are developed, which roughly resemble acini. In the latter, we have instead thin branching columns of cells, growing into the connective tissue stroma. All such neoplasms are exceedingly malignant. The development of cancer is sometimes accompanied by secretion, as shown by the escape of fluid from the nipple: cystic cavities are occasionally formed in cancer from the same cause.

This is the nearest approach to healthy function, that the gland attains under such pathological conditions.

In the *Adenomata* a much higher grade of organisation is attained. The newly formed gland tissue tends to be very like the normal, though as a rule it falls a little short of this high standard. The cells of the pathological acinus are generally massed in several layers, and they often form intra-acinous papillary projections. The lumen, as a rule, is absent or more or less occluded; and it has no connection with the excretory duct. Adenomata are generally incapable of producing normal secretion; but they are very prone to form serous and mucous fluids, which cause cysts to arise. Adenomata occasionally attain such a high degree of development that they resemble supernumerary mammæ and secrete milk. Such neoplasms never manifest infective qualities.

I will now pass to the *Parablastic* or *connective* tissues, and their corresponding pathological new formations.

At an early period of embryonic life, those parts of the body where connective substances will subsequently arise, are composed solely of closely aggregated parablastic cells These cells are nothing but small, rounded masses of nucleated protoplasm. Such is the embryonic parablastic tissue, whence any kind of connective substance may subsequently develop. Sometimes its cells become elongated or spindle-shaped, which seems to indicate tendency towards a higher stage of organisation. In man and the higher animals this tissue has normally only an embryonic existence; but a large group of pathological new formations—the *Sarcomata*—are composed entirely of it. In the arthropoda and mollusca it exists as a permanent tissue.

The first step in the evolution of the mature connective tissues, from this cellular embryonic structure is, that the cells become separated from one another, by the differentiation of intercellular substance from their protoplasm. A distinction thus arises between the cells which form and the intercellular substance which is formed. In the developmental process most of the embryonic cells are used up and converted into special tissue. In some instances the cells totally disappear, or only their nuclei are left. Usually, however, a considerable number persist, embedded in the intercellular substance, and still capable of active development. These are the cells which maintain the structure in its normal state, repair injuries, and, by reversion to an embryonic state of activity, originate the various parablastic neoplasms.

In this way the various kinds of *Sarcomata* arise. These neoplasms are composed almost entirely of cells, which are either rounded or spindle-shaped. The new formation seldom advances beyond this low grade of organisation, which is closely allied to that of embryonic parablastic tissue. All such neoplasms are more or less malignant.

In addition to the cellular embryonic form of connective tissue, the following varieties may be recognised: the gelatinous, fibrous, fatty, cartilaginous, and osseous. The factors upon which these differences depend are, the character of the constituent cells, the relation these bear to the intercellular substance, and the physico-chemical constitution of the latter. Modifications of these relations, whether arising during embryonic or post-embryonic life, cause transformations to occur between the different connective substances (*metaplasia*). Fibrous, cartilaginous, osseous, mucous, and adipose tissues are thus potentially convertible. This of itself is a weighty reason for admitting all these varieties into one group.

The formation of *gelatinous tissue* from the embryonic tissue, takes place by the differentiation of a soft, hyaline, intercellular substance containing mucin. In this substance the cells are sparsely embedded. They may be of rounded shape, when they are generally isolated; or stellate, when they usually communicate with one another by fine protoplasmic processes. This tissue is permanent in many of the lower animals, *e.g.*, the mollusca and medusæ. In the human embryo it is very abundant, existing wherever fibrous or fatty tissue will subsequently be formed. At birth it is found in the umbilical cord; and it is permanent in the vitreous humour of the eye. Pathological neoplasms are occasionally formed of this substance. These are the *Myxomata.*

Such neoplasms originate from the cells of fibrous or fatty tissue. These proliferate and produce embryonic parablastic tissue, whence the gelatinous tissue subsequently differentiates as in the embryo. Myxomata are a degree higher in organisation than Sarcomata. They represent immature fibro-fatty tissue; and are therefore lowly organised. Such neoplasms generally recur locally, and they not very unfrequently give rise to metastases. Gelatinous tissue in neoplasms is often derived by metaplasia from fibrous, fatty, and cartilaginous tissues; under these circumstances its presence is not of such serious import.

Fibrous tissue may be regarded as a further development of gelatinous tissue. In the process of evolution the diffluent, mucin-bearing, intercellular substance gradually fibrillates and solidifies, and is converted into gelatine. Many histogenists describe fibrous tissue as originating without a gelatinous transformation; the developing cells (fibro-blasts), are then said to arrange themselves into compact masses, and their protoplasm fibrillates. In the developmental process the cellular elements gradually waste and many disappear. However, a considerable number remain embedded in the fibrillar intercellular substance.

Abnormal proliferation of these cells is the first step in the formation of a *Fibroma* or fibrous tissue neoplasm; the embryonic parablastic tissue thus formed, passes into fibrous tissue as in the normal development. Fibrous tissue may be transformed by metaplasia into gelatinous tissue or bone. Neoplasms of this nature are highly organised, and perfectly innocent.

Fatty tissue is but a modification of ordinary areolar tissue, owing to the deposition of fat within its cells. The panniculus adiposus of the adult is, in the fœtus, a gelatinous tissue. In its evolution fatty tissue passes through the cellular and gelatinous stages; but, whilst the fibrillar stage is still incomplete, the cells form groups and fat accumulates within them. The accumulation of fat within the cell causes the nucleus, with the little remaining protoplasm, to be pressed against the cell wall. These cells, however, are still capable—under pathological conditions—of taking on fresh activity. Embryonic parablastic tissue is thus formed, which finally develops into fat, as in the normal evolution. Thus the *Lipomata* arise from normal fatty tissue. Sometimes, however, such neoplasms originate in connective tissue, which normally contains no fat. Lipomata are highly organised and non-malignant. Fatty tissue passes readily by metaplasia into gelatinous or fibrous tissue.

Cartilage is developed in the embryo from cellular

parablastic tissue, by the formation of firm, chondromatous, intercellular substance. At first this substance is scanty, and the tissue consists almost entirely of cells. This is the so-called cellular cartilage, which is permanent in the vertebrate notochord. As development proceeds, the cells multiply, and each new cellular group forms round it fresh cartilage. The intercellular substance, which is at first homogeneous, may retain this condition, as in hyaline cartilage; or it may undergo further differentiations, whence the fibrous and elastic varieties. Cartilage readily passes into gelatinous, fibrous, fatty, or osseous tissues. These metamorphoses may be effected directly, as when cartilage arises from the perichondrium, periosteum or marrow; or indirectly, through the intermediate stage of embryonic tissue, as in the formation of bone. Cartilage cells are, as a rule, spherical, and they are quite separate and detached from the intercellular substance in which they are embedded. In many selachians and cephalopods, however, a kind of cartilage is met with, in which the cells communicate freely by fine radiating processes. This form does not occur, as a normal production, in man; but it is sometimes found in pathological chondromatous new formations. Similarly calcified cartilage, which in man is normally only a transitory structure, is often met with in neoplasms; and in sharks it forms a persistent part of the skeleton.

Chondromata, as pathological cartilaginous new growths are called, mostly arise in connection with skeletal structures, or in the vicinity of certain glands. The matrix tissue may be pre-existing cartilage, perichondrium, periosteum, marrow, bone, or other connective substance. These neoplasms never develop directly from any adult structure; but the parablastic embryonic tissue is first formed, and the subsequent evolution is after the normal type. The degree of organisation attained by chondromata varies; some never progress beyond embryonic stages, and are as malignant as any sarcomata; others attain as high a degree of development as any of the adult varieties of cartilage, and are perfectly innocent.

All *Bone*, whether of embryonic or post-embryonic origin, is developed from the so-called osteogenetic tissue, which consists of small protoplasmic cells (*osteo-blasts*). These secrete the sclerogenous substance, which forms the essential constituent of bone. Each osteoblast contributes a zone of osseous matrix, in the centre of which it becomes embedded; at the same time it develops fine processes which communicate with other adjacent cells. Thus the so-called bone-corpuscles arise. The osseous substance is at first a soft fibrillar kind of connective tissue, which gradually hardens owing to impregnation with lime salts. The development of bone is often preceded by that of cartilage and fibrous tissue. In the latter case, ossification is effected by conversion of the connective tissue cells into osteoblasts, whilst the fibrous intercellular substance disintegrates. The development through cartilage is the more frequent, and it holds for most of the vertebrate skeleton. In this process the cartilage disappears partly by retrogressive changes, and partly by transformation into marrow and bone. In these cases metaplasia precedes the osteoblastic process. Formerly great importance was attached to these different modes of ossification; but in reality there is no essential difference between them, both processes being merely modifications of the method first described. Bones increase in thickness at the expense of the cellular layer of the periosteum, which originates the osteogenetic tissue, whence the new bone is developed.

Osteomata generally arise in connection with skeletal structures, such as periosteum, perichondrium, cartilage, bone, marrow and fibrous tissue. The cells of a part in which an osteoma is forming proliferate and develop into bone, as in the normal evolution, either directly or indirectly though fibrous tissue or cartilage.

The different varieties of parablastic neoplasms are connected by numerous intermediate transitional forms.

CHAPTER V.

ÆTIOLOGICAL.

"What interpretation we put on the facts of structure and function in each living body, depends entirely on our conception of the mode in which living bodies in general have originated."
SPENCER.

HITHERTO very little has been said with regard to the difficult subject of Ætiology; now this matter must be discussed.

The object of Ætiology, in its widest sense, is to ascertain the causes which determine the phenomena relative to organic form, function and distribution, by referring them to general laws. In this sense, I need hardly say, the science is in its infancy.

In the present chapter it will be my endeavour to indicate briefly the chief considerations, bearing on the ætiology of animal and vegetable neoplasms.

Notwithstanding the great advances recently made in biological science, in the majority of cases we are still unable to indicate the precise nature of the circumstances which determine particular variations in organisms.

Yet everyone who has investigated the nature of the local changes in neoplasms, must have been struck by the fact, that at every turn the phenomena met with have their counterparts in the normal processes of evolution. Between the laws which govern the evolution in the two cases, there is evidently no intrinsic difference—the normal is the type of the pathological. Hence, we may reasonably hope, that as the normal process becomes better understood, so our knowledge of the abnormal process will increase.

When we come to examine the linked chain of living things in the light of the doctrine of evolution, we are insensibly led to this important conclusion, that functional modifications in all cases go before structural changes, and are from beginning to end their determining cause.

As I have previously mentioned, in the development of every higher organism, there may be observed a twofold progress proceeding *pari passu*. On the one hand, there is a continuous perfecting of bodily structure by increasing histological and morphological differentiation, whence result the numerical increase and subdivision of tissues and organs; and, on the other hand, there is a continual transition from lower to higher types of development. These metamorphoses, as Baër has pointed out, indicate roughly the changes of structure undergone by ancestors. Each individual organism reproduces, in the rapid and short course of its own evolution, the most important morphological changes through which its long line of ancestors have passed.

The development of the individual from the fertilised germ (ontogeny), is, in fact, an epitome of that of the race (phylogeny), conditioned by the laws of heredity and adaptation. To Darwin is due the great merit of having placed the phenomena of heredity and adaptation in their true light. He it was who first showed how organic forms are changed, by the interaction of these two important physiological functions. So completely are they interwoven, that it is generally impossible to say how much of a given morphological change is attributable to the one, and how much to the other. Taken as a whole they stand in a certain opposition. Heredity is the cause of the stability of organisms, and adaptation of their modification. Heredity may be regarded as the conservative factor, ever tending to keep the descendants like the ancestors; whilst adaptation represents the progressive factor, striving to change the organism through the direct or indirect influence of the environment. According as the one or the other influence is paramount, the organism will remain constant or vary.

In every act of reproduction, a certain quantity

of protoplasm is transferred from the producing to the produced organism, and along with it, the molecular motion peculiar to the parental individuals.

The phenomena of *Heredity* are essentially dependent on this material continuity and partial identity, of the producing and produced organisms. If, instead of a succession of individuals thus produced, we substitute in imagination a single continuously existing individual, we shall at once see the relation in which an organism stands to the rest of the species; and how the succession of all living things is, as Goethe says, a linked chain.

Thus, the persistence of impressions (unconscious memory) in protoplasm is the property upon which, in ultimate analysis, the phenomena of heredity depend. Hence, every living thing produces new ones, each after its kind. It is in virtue of this property that, in Paget's words, "a mark once made on a particle of blood or tissue is not for years effaced from its successors."

For the initiation of the evolution of the higher organisms the blending of two cells, derived from different individuals is, as a rule, essential; both of these exert their influence in forming the offspring. Sometimes the male influence is seen to predominate in one direction, and the female in another. No definite rule can be laid down. Equal transmission by both sexes is the commonest, but sometimes characters are transferred exclusively to the sex in which they first appear.

It should be borne in mind that the actual product of the development of a fertilised germ by no means represents the full measure of its potentiality. The particular form evolved in any given case must be regarded as the resultant of the many fluctuating forces, inherent to the parental reproductive cells at the moment of impregnation.

Only a portion of the many varying tendencies inherited by the reproductive cells from their long line of ancestors are actually evolved in each generation; so that the transmission and development of qualities, though they usually go together, are in reality quite distinct powers. In virtue of this property it has been well said, that a man may best know himself in his relatives—parents, uncles, aunts, cousins, brothers, sisters, children; in them he will often see developed his own latent tendencies.

Similarly the secondary sexual characters of one sex lie latent in the opposite sex, ready to be developed under favourable conditions. In like manner qualities are transmitted through the earlier years of life, which are only subsequently developed. In accordance with the same law we frequently see qualities transmitted, in a dormant state, through numerous generations, and then suddenly redeveloped.

In some cases the number of intervening generations may be enormous. "In every organism," says Darwin, "we may feel assured that a host of lost characters lie ready to be evolved under favourable conditions." All of these wonderful phenomena are included under the term of atavism or reversion, which is an important subject in relation to the ætiology of morphological variations.

Not only can organisms transmit those qualities which they have inherited from their ancestors, but also those they have acquired during their own lifetime. New characters thus transmitted tend to appear in the offspring at the same age as they were first acquired by the ancestor; and when deviations from this rule occur the transmitted characters usually appear earlier. It has also been observed that local variations tend to appear in the offspring, at the same place as they first appeared in the ancestors. Moreover, variations which first appear, in either sex, at a *late* period of life, tend to be inherited by the same sex alone; whilst those which first appear early in life, in either sex, tend to be developed in both sexes. This is probably due to the fact that before puberty the sexes do not differ much in constitution.

Since no two parts of any organism are ever similarly conditioned, its individual germ cells can never be exactly alike, nor can the sperm cells which fertilise them; hence the products of such fertilised germs must always be more or less dissimilar, so that the transmission of variations is itself variable, in the direction both of increase and decrease. For instance, an individual quality in one parent, may be so counteracted by the influence of the other parent, that it may not appear in the offspring; or, not being so counteracted, the offspring may possess it in equal, greater, or less degree. Thus Spencer remarks:— "Amid countless different combinations of units derived from parents, and through them from ancestors, immediate and remote—amid the various conflicts in their slightly-different polarities, opposing and conspiring with each other in all ways and degrees—there will from time to time arise special proportions causing special deviations. From the general law of probabilities, it is inferable that while these involved influences, derived from many progenitors, must, on the average, obscure and partially neutralise one another; there must occasionally result such combinations of them as will produce considerable divergencies from average structures; and at rare intervals, such combinations as will produce very marked divergencies."

In many cases, we are unfortunately not acquainted with the precise conditions under which the transmission of acquired characters takes place. It is, however, known that certain of these qualities are much more easily transmitted than others, *e.g.*, phthisis, insanity, albinism, weeping willows, copper beeches, &c. In these instances the change transmitted seems to be chiefly a constitutional one.

It has also been observed that variations acquired during the lifetime of an individual organism are more likely to be transmitted to its descendants, the longer the causes of such variations have been in operation; and the longer newly-acquired variations have been transmitted by inheritance, the more certainly will they tend to reappear in future generations. On the other hand, variations generated by causes that act for a comparatively short time, generally prove to be evanescent, on account of the preponderating force of the previous ancestral balance. Hence organisms which have undergone adaptation sunder new conditions, tend to return to something like their original structure, when restored to their original conditions.

Such, then, are the chief facts with regard to heredity with which we are here concerned.

In ultimate analysis, the phenomena of *Adaptation* are traceable to the endless variability of protoplasm, owing to molecular changes excited in it by the unequal incidence of the ever-changing environment. Change of nutrition causes functional alteration, which ultimately develops into obvious morphological variation. But modification of function and its expression by gross structural change is a gradual process; hence adaptation is only noticeable in a long series of generations, whilst heredity can be recognised in every generation.

Organisms have become different either by immediate adaptations to unlike conditions of environment, or by mediate adaptations through survival of the fittest by natural selection, or by both; hence all variations must be regarded as adaptive modifications.

Inasmuch as natural selection acts only through and for the good of each individual organism, it is perfectly obvious that this principle cannot have brought about the development of neoplasms. On the contrary, so far as its influence has come into play at all, it has evidently tended to induce a return to the normal. In developing the doctrine of natural selection, which is peculiarly his own, I think there can be no doubt that Darwin failed to attach due weight to the action of the environment, independently of selection. A large number of variations appear to have arisen in this way, some of which are obviously disadvantageous to their modified possessors; whilst others are neither advantageous nor disadvantageous. With these, it appears to me, the various pathological new formations should be classed.

Such variations may be induced either by the

direct action of the environment on the whole organism, or only on certain parts of it; or indirectly, through some subtle alteration in the reproductive system, without there being any obvious change in the parental form.

In *direct* adaptation, the new incident forces call forth their concomitant structural changes during the lifetime of the affected organism; but such changes cannot be produced in this way without modifying the whole organism. For when one function is thus changed, others are indirectly changed, and so eventually all parts of the organism. In this way individuals become adapted to new conditions of life; and such modifications tend to be inherited. "Changes of habit" in plants, the various racial distinctions, and the effects of use and disuse may be instanced as examples of direct adaptation.

In *indirect* adaptation, it seems certain that either sexual element may be affected in such a manner as to cause modifications to arise in parts subsequently developed from it. Most monstrosities and malformations appear to have arisen in this way. The morbid impulse (molecular protoplasmic disturbance) thus generated, issues—through the parental reproductive cells—as disease in the offspring. Darwin has made us familiar with the extreme susceptibility of the reproductive system, and especially of the female reproductive system, to changed and abnormal conditions in the environment. He has pointed out that many wild animals in captivity are sterile, or even refuse to unite sexually; similarly, many plants in the cultivated state fail to produce good seed. The relations between the sexual organs and other parts of the body are evidently of the greatest importance.

It seems to me exceedingly probable, that we may attribute the origin of neoplasms to causes nearly akin to those which induce local malformations *per excessum.* The two classes of disease probably differ only in degree. I regard all of these new formations as attempts of the organism to adapt itself to changed, abnormal, or injurious conditions of life.

In the present imperfect state of our knowledge, it is impossible to fix upon the precise change required to bring about a given abnormality. We can only point out, in a general way, that such changes must be sought in the particular relation which each organism has with the environment, habitually in the individual or occasionally in the race. It must be the object of future research to determine the nature of the changes which cause particular variations.

For the present, therefore, we must content ourselves with such indications as can be gathered from the following considerations.

We have seen that variability of every kind is directly or indirectly caused, by the changed conditions of life to which each organism, and more especially its ancestors, have been exposed.

"Of all the causes which induce variability," says Darwin, "excess of food, whether or not changed in nature, is probably the most powerful." Thus domestic animals, bred for certain purposes, have been variously modified, according to the different quantities and qualities of the food supplied to them. Similarly, a soil possessing certain ingredients, in unusual quantity or of unusual quality, has caused the parts of plants supplied by such ingredients to develop abnormally.

In the case of animals, want of proper exercise has probably been an important factor in causing variability, as well as the effects of changed climate and general surroundings. These seem to be the chief reasons why domestic organisms are so much more liable to vary than natural species.

It has been ascertained that the influence of changed conditions accumulate, so that several generations must generally be exposed to the new conditions before any effect is visible. Thus slight changes often suffice to induce considerable variability. Once the process has commenced, the more disposed is the organism to vary still further.

When a new variation seems to come suddenly into existence, there is always under the surface an unbroken physiological evolutionary process. For instance, the sudden development of a neoplasm, in an otherwise healthy organism, is in reality the outcome of gradual and continuous changes in the evolution of the cells of the affected part, owing to changes in the nutrition of the tissues in that situation.

In determining the nature of particular variations, the constitution of the organism acted on seems to be a much more important element than the nature of the changed conditions. This is what may be inferred from the appearance of nearly similar modifications under different conditions, and of different modifications under nearly identical conditions.

Thus, under the same abnormal conditions, different individuals show their constitutional differences by being variously affected; and even in the same individual, similar abnormal conditions will now affect one part, and now another. We have ample proof of the great importance of the apparently slight structural differences that make up constitutional variations, in the extremely different susceptibility of certain races to various diseases. Hirsch's recent work on the geographical distribution of disease has been of great service in calling attention to these important facts, hitherto too much neglected.

What is known of bud variation points to the same conclusion. As Darwin insists, bud variation shows that variability may be quite independent of sexual reproduction and likewise of reversion. For instance, the sudden appearance of a moss-rose on a Provence rose cannot be attributed to reversion, for mossiness of the calyx has never been observed in any natural species. Here the aid of the so-called spontaneous variability has to be invoked, as a cloak for our ignorance. It is important to observe that bud variations are of much more frequent occurrence in plants that have been highly cultivated during a length of time, than in other and less highly cultivated plants ; and very few well-marked instances have been observed in plants living under strictly natural conditions.

I will now pass briefly in review the process of gall formation, which appears to me likely to throw some light on this subject ; for there is evidently a likeness between it and the process underlying other anatomical and pathological variations.

Galls arise in consequence of excessive local cell-growth and proliferation, excited by the virus instilled into wounds made by insects in depositing their ova. Considering the very small size of most gall-producing insects, the drop of poison instilled must usually be exceedingly minute. Hence, it probably acts only on a single cell, or on a very few cells, which, being abnormally stimulated, proliferate. The subsequent growth generally begins soon after the instillation of the poison, and it often progresses rapidly ; but sometimes there is long delay. To produce its appropriate effect, each virus requires to be placed in a suitable situation, where the conditions are favourable for its development.

It is remarkable that galls present definite form and character; moreover, each kind keeps as true to its type as any independent organism—each has, in fact, an individuality of its own.

Notwithstanding the close affinity between many groups of gall-producing insects, yet the number of different galls caused by them is very great. On the oaks of middle Europe alone, nearly a hundred different kinds have been described, all produced by species of gall-wasps. Hence we may infer that very slight differences in the nature of the poison causes widely different results; and that the nature of the fluid instilled is a much more important factor in determining the form of the gall, than the specific character of the tree acted upon. This is especially obvious when different kinds of virus are instilled into similar tissues, as when three or four different galls, produced by as many different insects, are formed on the same leaf.

In treating of this matter, Darwin remarks: "As the poisonous secretion of insects belonging to various orders has the special power of affecting the growth of various plants; as a slight difference in the nature of the poison suffices to produce widely different results; and lastly, as we know that the chemical compounds secreted by plants are eminently liable to be modified by changed conditions of life, we may believe it possible that various parts of a plant might be modified through the agency of its own altered secretions. With such facts before us, we need feel no surprise at the appearance of any modification in any organic being."

Shortly before his death Darwin began to experiment as to the possibility of producing galls artificially. In his recently published life, we are told that he made a considerable number of experiments, by injecting various reagents into the tissues of leaves, and with some slight indications of success.

We have seen that buds arise wherever there is an excess of nutritive materials, capable of being utilised for growth by the cells of the part. Under such circumstances buds may be formed wherever undifferentiated cells are present. The stimuli which determine the nutritive flux may be either intrinsic or extrinsic. In either case the result is the same; the protoplasm of the part thus excited to excessive growth and proliferation gives rise to collections of lowly organised cells. These, not being required for the building up of the parental structure, become superfluous, and are no longer retained directly under its controlling influence.

Groups of more or less completely emancipated cells thus arise, which, as they develop, assume an *individuality* of their own; the local processes cease to be subordinate to the specific hereditary tendency of the whole. Here we see growth exceeding a certain amount tending to the formation of new aggregates. Whether such centres of vital activity continue to develop in organic connection with the parental form, or after separation from it, is determined by the conditions of nutrition. If the cells from which a leaf-bearing shoot would normally spring are abnormally nourished, they may develop instead into a tumour or some other abnormal production.

It has been observed, in plants, that anything which arrests the growth of an organism at any part, or only interrupts the continuity of the cellular tissue, may determine bud formation. For example, it is well known that injuries often call forth a large number of adventitious buds. This is the effect of cutting or breaking off a branch, or of anything that interferes with the vegetation of the normal buds. Wounds of the bark are often followed by the same result. Many plants present in their bark little groups of roundish cells—lenticels—which originally lie beneath the epidermis, but when this wears away they are laid bare. As the growth of the part proceeds the bark is often rent at these places, so that the growing cellular tissue is freely exposed. At the edges of such rents adventitious buds are very liable to form, which may develop into tumours;

Every one acquainted with Hyde Park must have noticed the curious bossy outgrowths on the trunks of its plane trees; these frequently shed their bark. In the process small portions of the cellular parenchmya are exposed. At these spots smuts and other foreign substances lodge, which excite the cells to excessive growth and proliferation. Thus the deformity originates. I have observed, however, that whilst many trees suffer, all are not affected; and those that escape are the finest grown and most healthy looking. It seems as if the irritant was only able to cause these growths in the feebly constituted.

When a begonia leaf is placed in damp soil, and incisions are made across its nerves, buds spring from every incision; and as many fresh plants may be obtained in this way as the leaf has received

FIG. 28.
Buds formed along incisions into the base of a hyacinth bulb.

wounds. The Dutch bulb growers have very cleverly availed themselves of a similar property to propagate hyacinths. This they effect either by making two or three deep incisions into the base of the bulb —destroying the nascent flower stalk—when, after a short time, numerous small buds form along the edges of the cut surface (Fig. 28); or by scooping out the interior of the base of the bulb, and so leaving exposed the cut ends of the sheathing leaves, from which buds soon spring in great numbers. Leaves subjected to slight pressure and in process of decay, frequently develop buds in a similar way. Degrees of injury which fall far short of this, such as those produced by various mild irritants, as described by Waldenberg, cause local thickenings, owing to increased growth and proliferation of the cells of the part. In like manner, as we have seen, the various kinds of galls arise.

Similarly, in animals, we may attribute the genesis of all neoplasms to excessive activity of certain lowly organised cells of the part, determined by local excess of nutrition, the result of intrinsic or extrinsic stimuli.

Just as cells embedded in the stroma of an ovarium become ova by excessive growth at the expense of adjacent nutritive materials, which they divert from other cells; so we may infer that those cells which originate neoplasms, become different from their congeners in a similar way.

In all of these cases the intrinsic causes seem to play a more important part than the extrinsic causes.

As I have previously hinted, the initial cause of such variations is probably to be found in perversion of the secretion of the affected part. Thus Rindfleisch remarks: "By the change of substance in the tissues certain excretive substances are constantly being formed, which must generally be passed off from the tissues and organs in which they form, as well as from the fluids of the body at large, in order that the life of the individual may be undisturbed. These bodies have their chemical position between the organo-poietic bodies on the one hand, and the excreted matter of the kidneys, skin and lungs on the other. They are different for the different tissues, and on this difference depends the variety of the pathological new formations."

Such changes imply more or less alteration in the processes and structures; not of the affected part only, but throughout the whole body.

In no single instance has a neoplasm ever been caused intentionally by mechanical or chemical irritation. In animals, whose tissues are much more sensitive than plants, we have no certain evidence that external stimuli can even excite their cells to proliferation; they appear to cause only inflammation. Such stimuli are of themselves insufficient to set up cancer or tumour formation, in the tissues of a previously healthy individual.

On this subject De Morgan says:—"Out of a

hundred chimney sweeps or clay pipe smokers, a certain number may have chimney sweep's cancer, or lip cancer, the number varying perhaps according to the duration and extent of irritation. But the majority will not become cancerous, irritate how you will; and of the remainder, few probably would have cancer unless irritation were applied."

We may then conclude that whilst extrinsic causes sometimes determine cancer and tumour formation, they are by no means its necessary antecedents. Though a blow, a wound, or other severe injury, may in a few instances precipitate the formation of a neoplasm, as in cases of so-called acute traumatic malignancy, and though frequently repeated irritations of long duration and moderate intensity are often observed as the precursors of neoplasms, yet such stimuli can never excite the predisposition to the disease.

Thus we are driven to adopt Darwin's conclusion, that in each individual "the conditions of life play but a subordinate part in causing any particular modification; like that which a spark plays when a mass of combustibles bursts into flame, the nature of the flame depending on the combustible matter and not on the spark."

Such, then, is a very brief account of what I take to be the chief considerations bearing on this exceedingly difficult problem. Though they have hitherto been neglected, I believe that ere long their significance will be universally recognised; and I certainly expect that this recognition will eventually lead to the discovery of better modes of prevention and treatment.

CPSIA information can be obtained
at www.ICGtesting.com
Printed in the USA
BVHW041737010819
554890BV00012B/444/P